The Permaculture Garden

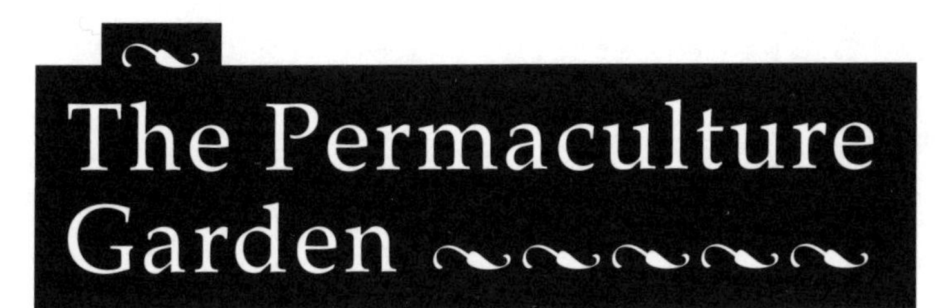

Graham Bell

Illustrated by Sarah Bunker

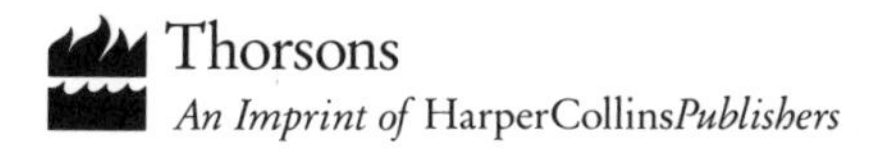

For Nancy and the
Glory of the Garden.

Thorsons
An Imprint of HarperCollins*Publishers*
77–85 Fulham Palace Road,
Hammersmith, London W6 8JB

First published by Thorsons 1994
Published by Thorsons 1995
3 4 5 6 7 8 9 10

A catalogue record for this book
is available from the British Library

ISBN 0 7225 2783 7

Illustrations by Sarah Bunker

Typeset in Palatino by HarperCollins
Printed in Great Britain by
Woolnough Bookbinding, Irthlingborough, Northants

Contents

Acknowledgements

I would like to thank the following gardeners for inspiration from their plots:

Eoin Cox, Jedburgh
Kate Cox, Holefield
Rod and Jane Everett, Middlewood
Phil and Anne Harris, Heatherslaw
Robert Hart, Rushbury
Ken Fern and all at Lower Penpol
John Manson, Greenlaw
Steve and Yvonne Page, Chez Forest
Emma and Bernard Planterose, Scourie
Clive Simms, Essendine
Owen Smith and Jill Jackson, Tai Madog
Tony Wrench, Wales

My apologies to the many more to whom I could have acknowledged my due debt.

I would like to thank the following co-teachers of Permaculture for what I have learned from them:

Chris Dixon, Sylvia Eagle, Patsy Garrard, Lea Harrison, Jane Hera, Joanna Jackson, Andy Langford, Ian Lillington, Bill Mollison, Stephen Nutt, Simon Pratt, George Sobol, Charlie Wannop, Patrick Whitefield.

In addition this work would not have been possible without the kind and loving support of Bob Clarke, Martin Hadshar, Steve Hoyle, Bruce Lowe, Julian Watson, Sandy Watson, Nancy Woodhead, Diana and Jay Woodhead.

Key to symbols used within the tables:

A	Annual
P	Perennial
B	Biennial
TP	Perennial—but tender
HH	Half-Hardy Annual

For more information on usage of individual plants, see my previous book *The Permaculture Way* (Thorsons, 1992).

All that matters is to be at one with the living God
To be a creature in the house of the God of Life
D.H. LAWRENCE (1885–1930), 'PAX'

Another Way of Gardening

I love gardens. When gardening, or just being in a garden, I feel truly alive, as if the garden connects me to a deep nourishing source of energy. Gardens remain a positive and vibrant joy in the centre of the lives of many people struggling to understand and cope with a pretty stressful and possibly crazy world.

This book is aimed at anyone with whom these thoughts resonate and is intended to encourage the many people who would like more out of their gardens, but find that they have limited time. It offers some approaches to gardening shared by those with a deep commitment to regeneration of both the Earth and the human spirit. 'Green' is one symbolic word used to describe this outlook. I'm not sure the label is too helpful, but it is an appropriate word to describe the levels of abundance you will achieve in your garden if you follow the advice contained within these pages.

When we spend time in the garden we are re-connecting ourselves to the living processes of the planet Earth, our home. Gardening is active leisure: not a 'pass-time' activity, but a creative enriching experience. Foolishly, we talk about 'being outside', as if the house were our 'real' place in the world, rather than the open air.

This sense of disconnection is a direct result of a series of revolutions; agricultural, industrial, and most recently, that of information technology. Before these changes tore society apart the majority of people passed the majority of their time with a direct connection to and a keen interest in their natural environment.

This book proposes another way of gardening, one which re-establishes our sense of belonging to the world, of being an intimate part of the living cosmos, rather than separate from it. It doesn't suppose that we're all 'going back to nature' (we never left it), or even that we have to give up our day jobs. Nor does it propose a lot of hard work. It *does* reveal a way in which ordinary people can do their bit to re-green the planet.

Large sentiments for so humble a pursuit as gardening, perhaps. Yet we stand at a crucial time of choice in human affairs. Either we continue to regard our powers of control over nature as infinite, and the Earth a bottomless well of resources for our consumption, or we face up to the degradation of environment and human soul which is globally evident, and decide to accept responsibility. In the face of impending global disaster most of us feel overwhelmed—but our back yards offer us an immediate avenue for practical planet care.

Good-quality land is increasingly scarce. World population is growing. Oil is running out. We could take these as a doom-and-gloom scenario, or, if you take my tip, you'll garden. Growing your own food, making tiny part-acres into vibrant mini-forests, is a direct and positive reaction to set things right.

In Kerala in India (a state jam full of gardens), a half-acre is seen as a farm which should feed a family. The average back garden may be smaller than this, but can still be highly productive, and in doing so take no more than an hour or two a week by using design methods which minimize effort.

When we garden we make contact through all our senses with the vagaries of weather, the changing moods and demands of the seasons, even the relative movements of the stars and the phases of the moon. In so doing we re-develop within ourselves an understanding of the quality of the fertile processes which make our gardens abundant, and feel ourselves again as part of the continuum of life, rather than separate observers.

This process of designing our lives from observations of nature based on seeing ourselves as part of nature has been called Permaculture, derived from the words *Perm*anent Agri*culture*, but also implying permanent culture. My previous book, *The Permaculture Way* (Thorsons 1992) provides an introduction to that design methodology in general terms. In this book we examine specifically how Permaculture can be applied in gardening. Later in this introduction we shall look at the whole question of culture, and how strongly gardening features in our cultural inheritance.

What *is* Permaculture?

A short answer would be that it is the art of the possible. Environmentalism was brought to public attention in the 1960s around 'conservation' issues. Permaculture seeks to go beyond that to *regenerative* policies which will restore global environments over time in a living changing way, rather than simply halting the further destruction of small parts of the planet, and preserving them as museum pieces.

From small beginnings it has become a major body of knowledge about how to design the places where people live and work.

Only that which is sustainable can build permanent societies. Our present unstable politics, land use and lack of care for people can only lead to change. It is a change that will bring us back to enduring societies and is one that must be based on permanent agriculture.

Why do we need permanent agriculture, and what would it look like if we ever saw it? We are rapidly learning that the price of our present high level of consumption is damage to the global environment on a massive scale.

Anyone who watches television or reads newspapers will have seen plentiful evidence of this presented over the last few years. Be it holes in the ozone layer, pollution, wars over oil or global warming, all threaten our future food supplies.

Permaculture is a term coined by an Australian, Bill Mollison, in 1978, when he wrote a book called *Permaculture One* with David Holmgren. The concept is based on many years' observation of natural systems. The peak example in a temperate climate might be deciduous forest.

In true wilderness (of which there is virtually none left in Europe), a forest is a system of plant cover which is self-regenerating and indefinitely sustainable. It is a system that functions in five dimensions: the two horizontal dimensions, the vertical dimension, and the added dimension of time, its crowning glory being the fifth dimension of relationships. Each of these 'directions' maximizes spatial use, adding to the productivity of the whole.

Mechanized monocultural agriculture is very two dimensional. In contrast, our example of the forest offers a large range of possibilities for life from the deepest root to the tallest treetop. The tree itself changes through the seasons, so that in early spring bulbous plants flourish before dense leaf cover cuts out the sunlight. Even daily changes offer successions of opportunity for different mammals, birds and insects to browse and carry out their other life functions. The tree and other life forms flourish, not in isolation, but by virtue of their many beneficial relationships. The tree's roots draw up nutrients, not just by physics and chemistry, but in intimate associations with all manner of soil life, from earthworms to bacteria. In turn, it may not be able to continue its chain of life without insects to pollinate its flowers, and mammals and birds to spread its seeds. This is just a tiny part of the complex web of relationships of a single tree.

By observing natural cycles, people practising Permaculture have developed these principles into strategies that enable anyone to build systems anywhere on Earth, which:

- are high yielding, regenerative and sustaining
- require minimum effort for maximum output
- are ethical, caring for the land and caring for people
- generate surplus for sharing.

New houses, new furniture, new
streets, new clothes, new sheets,
everything new and machine made
sucks life out of us
and makes us cold, makes us lifeless
the more we have

D.H. LAWRENCE (1885–1930), 'NEW HOUSES, NEW CLOTHES'

Adopting Permaculture in your garden could be the first step towards limiting your personal consumption and planning your life to become more creative as time goes by.

Controlling our consumption means looking very hard at our inputs and outputs. In the forest no-one brings in manure by the truck load, or takes away polythene bags of unwanted woody cuttings. All the needs of the forest have to be supplied from within itself, and its outputs must be used up by other elements of the system. If earthworms and other soil flora and fauna did not break down and digest fallen leaves the forest would become polluted by them.

Pollution is just unwanted resources. So in Permaculture we are learning to make appropriate placement of all inputs and outputs, aiming to reduce work and minimize pollution.

Your sustainable garden will be a very visible testimony to the art of the possible. It will also be a living example of a design approach which can be applied to all aspects of your community. Gardening is rooted deeply in human culture as a practical symbol of all that is life-enhancing. The garden is nature understood, and disciplined through care.

A Vision of Global Gardening

I remember people coming to my mother's yard to be given cuttings from her flowers; I hear again the praise showered on her because whatever rocky soil she landed on, she turned into a garden...She is involved in work her soul must have. Ordering the universe in the image of her personal conception of Beauty.

ALICE WALKER, *IN SEARCH OF OUR MOTHERS' GARDENS*

For many black women in the Southern United States gardening was the sole allowable expression of their inner creativity. I would suggest that this outlet for the human psyche is common to many more people. It is not an accident that Christianity has our creation take place in the Garden of Eden, and that when Adam and Eve are banished from the sight of God it means they must leave the garden. Or that, according to the Koran, after death the just will rest with God in a beautiful garden.

The word 'Paradise' itself is derived from the old Persian *Pairidaeza*, meaning a walled garden. The image is used often in literature, especially allegory, as representing the human soul. In *The Allegory of Love* C.S. Lewis comments:

Do not let us be deceived by the allegorical form. That...does not mean that the author is talking about non-entities, but that he is talking about the inner world—talking, in fact, about the realities he knows best.

Another common strand is that many people have a 'garden dream', a single or repetitive dream in which they are in a secret garden (usually with their siblings), which is in some way a safe, calm and happy place. There is a door to the garden, and the dream is usually without any action. This is thought to be a memory of life in the womb, translated into terms intelligible to the post-birth subconscious.

The way in which our consciousness across many cultures has made such a deep commitment to The Garden as an image is only surpassed by the actual practice of gardening as a human bond. It is one of the few activities that people share throughout the world and, as such, if we see ourselves as gardeners first and foremost, reveals our common humanity. From such bonds are peaceful and creative futures made.

Some final reasons for the garden as a starting point for environmental regeneration are:

- local: gardens are usually next to where we live
- personal: their cultivation can be done individually
- inviting: gardens invite sharing and participation
- achievable: the skills needed are easily acquired.

And so gardening is an activity available to the great majority of people. It can even be practised in window boxes and on balconies, and without soil at all. In my vision of a peaceful and abundant future for the planet, I see the virtual dismantling of large-scale agriculture and monoculture forestry, and in its place autonomous regions of forest gardeners, making fewer demands of the planet and of themselves than we do now, living a simpler life, with adequate food, clothing and housing, and probably a great deal more leisure than we have now.

Time to play with our children, be with our lovers, time to walk in the countryside or simply to sit and admire the ever-changing prospect of our gardens.

Every person needs to have a piece of garden, however small, to keep them in touch with the earth and therefore with something deeper in themselves...

CARL JUNG (1875–1961)

I hope this book adds dimensions of pleasure and productivity to your life in the garden. We need not live out our lives in fear of war, pestilence or environmental destruction. Mother Nature is showing we need to make changes. We might see these will actually make our lives more pleasant and fruitful. In short, our gardens might be places to mend our souls, and experience profound joy in living. It is in that spirit that I invite you to enter the *Permaculture Garden...*

Graham Bell

Coldstream, May Day 1993

one

What is a Garden?

Sunlight,
Three marigolds,
And a dusky, purple poppy-pod–
Out of these I made a beautiful world

AMY LOWELL (1874–1925), 'FUGITIVE'

To design anything consciously we must understand the importance it has in our lives. In gardening people often do what is fashionable, rather than what would suit them best. There are some easy questions we can ask ourselves to make sure the garden works for us, rather than the other way round.

Beauty and Use

Everyone wants a beautiful garden. Pleasure to the senses is one thing to one person, and something different to another. Yet we all know that feeling of going into a garden and feeling we're somewhere *special*–even if it's not always our own plot! When beauty is in the eye of the beholder, how do we decide what makes a garden aesthetically pleasing?

There's a school of thought that sustainable should mean 'useful', as if fripperies such as 'beauty' have no place in the scheme of things. Beauty is, however, a useful commodity. It is, if you like, food for the soul. In a garden that aims to emulate nature, by producing the sort of bewildering profusion with which wilderness delights the senses, there may be missing some of the traits that make many conventionally 'beautiful' gardens acceptable.

Principally there is the obsession with 'tidiness'. If you follow the ideas in this book you will have little interest in bare weeded soils, nor will you enjoy the sight of straight regimented rows of even-sized plants laid out like so many formal wallpaper designs. Order will be built of subtler patterns. In time what happens is that the prolific and random abundance of your natural garden will rival any formal plot for its ability to delight the senses.

Sustainable gardening is a controlled

replay of the fecundity of natural systems. So beauty becomes the riot of colour, shape, scent, texture, light, shade, height and contrast which form the fertile fabric of unfettered nature itself. Every plant has aesthetic beauty in its own right, from the least desirable weed to the most fragile hothouse exotic. The beauty of any part of the Permaculture garden is in its healthy development.

It comes out in the products themselves, not just in taste and the satisfaction of an abundant harvest of food, but right down to how you lay out a meal attractively.

We find things beautiful when they do what we intended for them in the first place. There is also the beauty of uncontrolled happenings in the garden. An unexpected weed may delight us because it teaches something about the processes of nature functioning under her own steam.

Work and Play ~

The back yard is a place of both work and play. That, as adult gardeners, we all know. Most gardens are frequently play areas for children. Nearly every garden is festooned with drying washing at least weekly. They're also the places where we eat out in fine weather, park and mend our bicycles, store the garbage, keep bits of old junk which might be useful one day, spend intimate moments with our loved ones, relax in retirement, or any other number of activities which don't immediately fit in with hybrid tea roses and the mixed herbaceous border.

Design and maintain your garden as a real place, not the fictitious elegance of a television gardening programme. We can all be made to feel inferior in the face of some carefully tended plot which has just received two weeks' intensive manicuring before the cameras turned up. Reality is different. Let's all loosen up and accept our plots as they actually are. At the end of the day if you *enjoy* your garden, you can't be going far wrong.

Food ~

People eat food, they do not ingest nutrients.

MAGNUS PYKE, *FOOD & SOCIETY*, 1968

Our first need in settling down to farm was to grow our own food. The earliest history of gardening is something we know only by conjecture. We assume that at some point in time people moved from being hunter-gatherers to agriculturalists. Literally, we cultivated the land.

Enough food of the right quality is essential to health. We are all responsible for eating a sensible mixture of food which meets our needs for energy for living, growth and renewal of the body, and for satisfying and pleasant meal times. We can choose to eat in a healthy way, but how we define that will vary from person to person. Raw slugs, mashed swede, grass seed and dandelions in apple juice would be a perfectly nutritious meal with a good balance of protein, carbohydrate, fibre, fat and vitamins, but one which most people would find hard to stomach.

We eat, and do many other of life's practices, not merely to avoid pain (hunger) but to have positive pleasurable experiences.

Growing our own food in the garden brings us directly to an appreciation of the efforts and rewards of eating fresh produce. We become conscious not just of a delicious bowl of new potatoes, for instance, but of the very weather which influenced their growth, of the nature of the soil which transubstantiates sunlight and water into our next dinner, and of our inherited place as sons and daughters of the earth. For we have as much right to the pleasures of the garden as the wild birds and insects who enjoy its liberty in such a seemingly carefree way, and all too often seem to think we grow food solely for their benefit.

The point of growing our own food is not merely to pay good attention to the health of our bodies and spirits through the joy of eating our own produce.

It's no problem to enter our local supermarket and scoff anything from kumquats to starfruit, yard-long beans to beef tomatoes, and that on just about any day of the year. I have seen 'organic' tomatoes on sale when the snow lay half a metre deep outside. Organic? How? It's not just that eating out of season leads to a profligate waste of energy in shipping and preserving the produce half-way round the globe. It's also that in many countries where these export cash-crops are produced, they are being grown at the expense of food for local people.

Every fruit or vegetable is fertility lifted from the land. If we are not returning the crop wastes to the land from which the food came then we are in effect part of a worldwide system of erosion.

Many areas of Britain have seen the contraction of the market gardening industry, where good food was grown to be fresh for local consumption. As urban expansion has taken over good growing land we still have half a million hectares of back gardens ready to leap into productivity. Part of the problem has always been that it seemed like hard work, and that growing vegetables meant you couldn't have an attractive garden. As we'll see, this doesn't have to be true.

We don't need to waste any time feeling guilty about the fact that we *occasionally* eat imported food, if we are moving in the direction of growing more of our own. A major point of gardening in a sustainable way is to produce food to eat ourselves.

Fibre and Fuel

Once people grew their own clothing and heating materials. Before the industrial age regions were self-reliant in fibre and fuel stocks to a much greater extent than today. There was little spare energy in the marginal economies of land-based living to waste on shipping these commodities around.

The sustainable Garden: before...

and after.

Only the wealthy could afford to do so on an appreciable scale.

Every village had its own flax fields for linen production. The sheep became the wonder beast of the middle-ages because of the diversity of its products, and in particular the high value of its wool. In practice many people had to look creatively for their fibre needs to make clothing and other artefacts, such as rope. Hemp and flax pollen are commonly found in mediaeval archaeological sites, indicating their usage as fibre plants. In remoter parts of Scotland bracken was used to plait ropes, and in many countries nettles were a common source of cloth fibre. Resources which are now regarded as 'by-products' such as tree bark were highly valued for specific purposes.

The same was true with fuel. Much timber which is now carelessly taken as firewood would have been considered of constructional quality in an age when people built timber structures and furniture in accordance with the shape and grain of the tree, rather than obeying the straight line dictates of the mechanical bench saw. Obtaining firewood was often a hard-won process of gleaning scraps of wind-blown twig, rather than chain sawing and splitting mature hardwood. In China the practice of saving every last scrap of vegetable material for some purpose or other meant that people used cabbage stalks as firelighters. The idea of a garden bonfire to 'burn waste' would have been a complete anathema to such folk.

In the sustainable garden we will return to seeing how much of our needs in these areas we can meet from the backyard. It doesn't mean we'll all be wearing nettle frocks, and eating weed soup cooked on a cabbage stalk smoulder (I hope). But we may take great pleasure in discovering just how much we thought we needed that we don't, and how much of what we do need is easily available to us from a small space. Thrift is a word that went out of currency when wartime austerity gave way to 'you've never had it so good'. There can't be a family that hasn't felt the effects of the recent economic recession to some degree. It's a timely reminder that thrift is actually a cardinal virtue, which it can be a necessity (and a pleasure!) to practice.

Healthy Soil ~

Great civilisations have almost invariably had good soils as one of their chief natural resources.

NYLE C. BRADY, *THE NATURE AND PROPERTIES OF SOILS*, 5TH EDITION, 1974

As all organic growers know, the end result of any growing process simply reflects the health of the system; holistic health care for the land, if you like. Healthy soil growing a balanced harvest offers us a much greater possibility of good yields of high nutritional value. Styles of growing that ignore the need to keep the system healthy are forms of mining. We rob the soil of its nutrients at our peril.

So the first objective of a sustainable garden is to look to the health of the land itself. If our gardening activity is building rich, well-structured soil we don't need to worry about crops. They will come as a natural yield of an abundant system.

Looking for permanence in any kind of horticulture means building a garden that is as self-maintaining as possible. Where you are changing from a garden managed with chemicals, or from one where the natural life of the soil has been damaged by over-cropping or compulsive digging, it may take time and energy to build a sustainable system. At the end of the day building living soil remains the main object.

Gardens for People ~

People need gardens, and gardens need people. The garden shouldn't be designed as a museum for plants, or as just a decoration attached to the house. It should be somewhere where people can feel comfortable. It shouldn't be all about 'doing': 'being' is just as important. We're often all so busy that maybe we should be called human-doings, not human-beings!

Think of how the garden might be arranged so that there's maximum human delight. Garden literature often shows plants without people. Rather

Magical garden bower: gardens need people.

than copying the perfect photos of the latest coffee-table book, imagine your garden filled with your friends and family having a good time. How would it look then?

Gardening is therapeutic. It can be carried out on your own ground, at home, preferably as near the back door as possible, and it is within your own power. That is not to say you will have it all your own way. Was ever life thus? In accordance with most earthly activity nature will always have a few surprises to pull on you. But at the end of the day gardening would be hard put ever to be harmful, and it is a way of taking responsibility for your own needs immediately.

A garden can be created in a day (yes, really!) It can be worked for five or ten minutes at a time. It can be admired. It can be shown off. It can be shared with others. It can be passed on when you move home. And it is always alive.

Environmental concern can generate many fearful images. The constructive act of creating something irrefutably alive is one of the most positive responses available to us.

> *(There)...lies the world-wide dream of the happy garden...the earthly Paradise...a system of conduit pipes which taps the deep unfailing sources of poetry in the mind of the folk, and convey their refreshment to lips which could not otherwise have found it.*
>
> C.S. LEWIS, *THE ALLEGORY OF LOVE*, 1936

Because we retreat to the solace of our garden does not mean we are turning our backs on the world. Far from it. The healing action of working in our gardens makes us much better able to deal with the stresses of our daily lives.

Gardening is also an activity which is shared in common by cultures throughout the world. The word 'peasant' is tainted in English with implications of class prejudice. In France a *paysan* is simply an inhabitant of the land (*paysage*), which in a nation in love with its diet implies a rich and fruitful sense of connectedness. Small-scale land use has survived there better than in the UK, where large commercial farm-as-business is the order of the day. Gardening is the peasant way of life, born in necessity, but grown through loving labour to a form of meditation, a communal bond, and a seasonal rhythm transcending the clockwork haste of a global society driven by technocracy.

The fact is that gardening brings us together and liberates the soul. Green fingers are not determined by class or wealth, and the joy of a garden returns the Countess and the farm labourer to their common humanity.

Table 1 Useful Ornamentals

Plants can be useful and attractive

a) Some perennials for human, wildlife and soil food

Bamboo	*Arundaria spp*	P Edible shoots/hedging
Barberry	*Berberis spp*	P Berries/hedging
Broom	*Sarothamnus scoparius*	P Edible flowers/nitrogen fixer
Buddelia	*Buddelia davidii*	P Butterfly bush
Heather	*Calluna vulgaris*	P Edible flowers/bee fodder
Ornamental quince	*Chaenomeles japonica*	P Fruits for jam/train as climber
Bladder senna	*Colutea arborescens*	P Nitrogen fixing
Hawthorn	*Crataegus spp*	P Edible buds/flowers/fruit/hedging
Bell heather	*Erica cinerea*	P Edible flowers/bee fodder
Bronze fennel	*Foeniculum vulgare dulce*	P Herb, all edible
Hibiscus	*Hibiscus spp*	P Edible leaf/flower/tea
Honeysuckle	*Lonicera spp*	P Edible flower/climber
Mahonia	*Mahonia japonica*	P Edible berries/hedging
Rose	*Rosa rugosa*	P Edible fruit/flower/hedging
Comfrey	*Symphytum officinale*	P Edible flower/green manure
Gorse	*Ulex europaeus*	P Edible flowers/nitrogen fixing/hedging

Do not eat flowers of Spanish broom Spartium juncium, which are poisonous.

b) Ornamental leaf vegetables

Egyptian onions	*Allium cepa var. viviprium*	P Has baby onions on top!
Amaranth	*Amaranthus gangeticus*	A Spinach substitute
Gold orach	*Atriplex hortensis*	A Salad/green vegetable
Red orach	*Atriplex hortensis rubra*	A Salad/green vegetable
Ruby/rainbow chard	*Beta vulgaris cicla*	A Salad/green vegetable
Romanesco broccoli	*Brassica oleracea italica*	A Spiralled heads of broccoli
Ornamental cabbage	*Brassica oleracea ssp*	A Edible
Frizzy endive	*Endivia riccia*	A Salad/Beautiful flowers
Ox-eye daisy	*Leucanthemum vulgare*	P Both young leaves and flowers edible
Red lettuce	*Lactuca sativa*	A Salad
Nasturtium	*Trapaeolum major*	A Salad/seeds as capers
Hamburg parsley	*Petroselinum crispum*	B Salad/herb

c) Edible flowers

Chives	*Allium spp*	P Salad
Hollyhock	*Althea rosea*	P Salad
Alkanet	*Anchusa officinalis*	B Salad

continued overleaf

Dill	*Anethum graveolens*	A Salad
Daisy	*Bellis perennis*	P Salad
Borage	*Borago officinalis*	A Salad/fritters or in fruit cup
Mustard etc.	*Brassica spp*	A Salad/stir fry
Marigold	*Calendula officinalis*	A Salad
Edible chrysanthemum	*Chrysanthemum coronarium*	A Salad/stir fry
Chicory	*Chicorium intybus*	P Salad
Pinks	*Dianthus spp*	P Salad
Meadowsweet	*Filipendula ulmaria*	P Salad
Lady's bedstraw	*Galium verum*	P Rennet substitute/summer drink
Speedwell	*Veronica spp*	P Infuse to make tea
Gladioli	*Gladiolus spp*	P Salad
Day lily	*Hemerocallis spp*	P Salad
Jasmine	*Jasminum spp*	P Salad
White dead nettle	*Lamium album*	P Salad
Red dead nettle	*Lamium purpureum*	P Salad
Lavender	*Lavendula spp*	P Salad
Crab apple	*Malus malus*	P Salad/Fritters
Common mallow	*Malva sylvestris*	P Salad
Chamomile	*Matricaria recutita*	A Salad/Tea
Lemon balm	*Melissa officinalis*	P Salad/Tea
Sweet bergamot	*Monarda didyma*	P Salad
Marjoram	*Origanum marjorana*	P Salad
Passion flower	*Passiflora caerulea*	P Salad
Pelargonium	*Pelargonium spp*	TP Salad
Butterbur	*Petasites vulgaris*	P Fry in butter
Petunia	*Petunia hybrida*	P Salad
Cowslip	*Primula veris*	P Salad/Wine
Primrose	*Primula vulgaris*	P Salad
Rose	*Rosa spp*	P Salad/Tea
Rosemary	*Rosmarinus spp*	P Salad
Clary sage	*Salvia sclarea*	A Salad
Painted sage	*Salvia horminum*	A Salad
Sage	*Salvia officinalis*	P Salad/Tea
Elderflower	*Sambucus nigra*	P Fritters/Wine
Scorzonera	*Scorzonera hispanica*	P Salad
Alexanders	*Smyrnium olusatrum*	P Salad
Golden Rod	*Solidago spp*	P Salad/Tea
Lilac	*Syringa vulgaris*	P Salad
Dandelion	*Taraxacum officinale*	P Salad/Wine

Thyme	*Thymus spp*	P Salad
Lime	*Tilia europaea*	P Tea
Salsify	*Tragopogon porrifolius*	B Salad
Nasturtium	*Trapaeolum majus*	A Salad
Red clover	*Trifolium pratense*	P Salad
Coltsfoot	*Tussilago farfara*	P Wine
Mullein	*Verbascum spp*	P Salad
Violet	*Viola odorata*	P Salad
Garden pansy	*Viola wittrockiana*	P Salad
Heartsease	*Viola tricolor*	P Salad

Table 2 Edible Plants

A huge range of food plants will grow in temperate climate gardens. Here is a range of possible domestic plant crops

Onion	*Allium cepa*	A
Welsh onion	*Allium fistulosum*	P
Leeks	*Allium porrum*	A
Garlic	*Allium sativum*	A
Chives	*Allium schoenoprasum*	P
Amaranth	*Amaranthus gangeticus*	A
Angelica	*Angelica archangelica*	B
Chervil	*Anthriscus cerefolium*	A
Celery	*Apium graveolens var dulce*	A
Celeriac	*Apium graveolens var rapaceum*	A
Asparagus	*Asparagus officinale*	P
Oats	*Avena sativa*	A
Bamboo shoots	*Various spp*	P
Beetroot	*Beta vulgaris*	A
Swiss chard	*Beta vulgaris var flavescens*	A
Spinach beet	*Beta vulgaris var vulgaris*	P
Mustard	*Brassica alba*	A
Pak choi	*Brassica campestris var chinensis*	A
Pe-tsai	*Brassica campestris var pekinensis*	A
Mustard spinach	*Brassica campestris var perviridis*	A
Turnip	*Brassica campestris var rapa*	A
Turnip greens	*Brassica campestris ssp*	A

Mustard greens	*Brassica juncea ssp juncea*	A
Kohlrabi	*Brassica oleracea convar acephala var gongulodes*	A
Kale	*Brassica oleracea convar acephala var sabellica*	
Cauliflower	*Brassica oleracea convar botrytis var botrytis*	A
Calabrese	*Brassica oleracea convar botrytis var italica*	A
White cabbage	*Brassica oleracea convar capitata var alba*	A
Red cabbage	*Brassica oleracea convar capitata var rubra*	A
Savoy cabbage	*Brassica oleracea convar capitata var sabauda*	A
Brussels sprouts	*Brassica oleracea gemmifera*	A
Marigold	*Calendula officinalis*	A
Sweet pepper	*Capsicum annuum*	A
Garland chrysanthemum	*Chrysanthemum coronarium*	A
Endive	*Cichorium endivia*	A
Chicory	*Cichorium intybus var foliosum*	A
Winter purslane	*Claytonia perfoliata*	A
Coriander	*Coriandrum sativum*	A
Japanese parsley	*Cryptotaenia japonica*	A
Gherkin	*Cucumis sativus*	A
Cucumber	*Cucumis sativus*	A
Pumpkin/winter squash	*Cucurbita maxima*	A
Pumpkin/winter squash	*Cucurbita moschata*	A
Marrow/courgette	*Cucurbita pepo*	A
Cardoon	*Cynara cardunculus*	P
Globe artichoke	*Cynara scolymus*	P
Carrot	*Daucus carota*	A
Salad rocket	*Eruca sativa*	P
Buckwheat	*Fagopyrum spp*	A
Florence fennel	*Foeniculum vulgare var azoricum*	P
Jerusalem artichoke	*Helianthus tuberosus*	P
Barley	*Hordeum distichon*	A
Celtuce	*Lactuca sativa var angustana*	A
Cabbage lettuce	*Lactuca sativa var capitata*	A
Loose leaf lettuce	*Lactuca sativa var crispa*	A
Cos lettuce	*Lactuca sativa var longifolia*	A
Garden cress	*Lepidium sativum*	A
Lovage	*Levisticum officinale*	P
Asparagus pea	*Lotus edulis*	A
Tomato	*Lycopersicon esculenta*	A
Alfalfa/lucerne	*Medicago sativa*	P
Ice plant	*Mesembryanthemum crystallinum*	P

Parsnip	*Pastinaca sativa*	A
Parsley	*Petroselinum crispum spp*	P
Hamburg parsley	*Petroselinum crispum spp tuberosum*	P
Runner bean	*Phaseolus coccineus*	A
Dwarf bean	*Phaseolus vulgaris*	A
Peas	*Pisum sativum spp*	A
Purslane	*Portulaca oleracea*	A
Black radish	*Raphanus sativa var niger*	A
Radish	*Raphanus sativa var sativa*	A
Rhubarb	*Rheum rhabarbarum*	P
Scorzonera	Scorzonera hispanica	A
Rye	*Secale cereale*	A
Aubergine	*Solanum melongena*	A
Potato	*Solanum tuberosum*	A
Spinach	*Spinacea oleracea*	A
Japanese artichoke	*Stachys sieboldii*	P
Dandelion	*Taraxacum officinale*	P
New Zealand spinach	*Tetragonia tetragonioides*	P
Nasturtium	*Trapeolum majus*	A
Bread wheat	*Triticum aestivum*	A
Cone wheat	*Triticum turgidum*	A
Corn salad	*Valeriana locusta*	A
Broad beans	*Vicia faba*	A

Table 3 Some Fibre Plants

Fibre plants for cool climates.
Trees and most shrubs can provide fibrefor different purposes

Bamboos	*Arundaria* spp etc.	P
Soft rush	*Juncus effusus*	P
Flax	*Linum usitatissimum*	A
New Zealand flax	*Phormium tenax*	TP
Bracken	*Pteridium aquilinum*	P
Wheatstraw	*Triticum* spp	A
Reeds	*Typha* spp etc.	P
Nettles	*Urtica dioica*	P
Yucca	*Yucca* spp	P

two

Planning your Garden

The heavens themselves, the planets, and this centre,
Observe degree, priority and place,
Insisture, course, proportion, season, form,
Office, and custom, in all line of order.

WILLIAM SHAKESPEARE (1564-1616) TROILUS & CRESSIDA

The sustainable garden is modelled on nature. Nature works upon the basis of patterns in many dimensions; it is no accident that rivers curve, or that bees build honeycomb in hexagons. Learning from observation of nature shows us that patterns are the way to build strong and enduring systems.

Flat Space and Edge

Let's start by thinking of flat two-dimensional areas–as if the garden were a map. We can see why nature finds strength in creating building blocks which are patterned. Consider a garden pond. Let us say it has an area of ten square metres, and let us make it round. How long is the edge? About ten metres, I make it. Look at the suggested designs for ponds overleaf. They all have the same surface area, yet they have remarkably different lengths of edge.

For another example, imagine a jigsaw made of squares. It has no strength and falls apart! A real jigsaw works because it has lots of edge.

When planning the garden the idea must be extended to every aspect of time and space.

Ecology is the study of how all nature is made up of different species living together. It emphasizes that nothing stands alone–we all thrive by interdependence with our neighbours.

The most fruitful places in nature are where different ecologies meet. This is because it is where there is most opportunity for swapping energy usefully. So at the edge of a woodland and a meadow we have the species which love sun, and those which seek shade, and those which need the special climate of that particular situation.

Within the depths of the woodland,

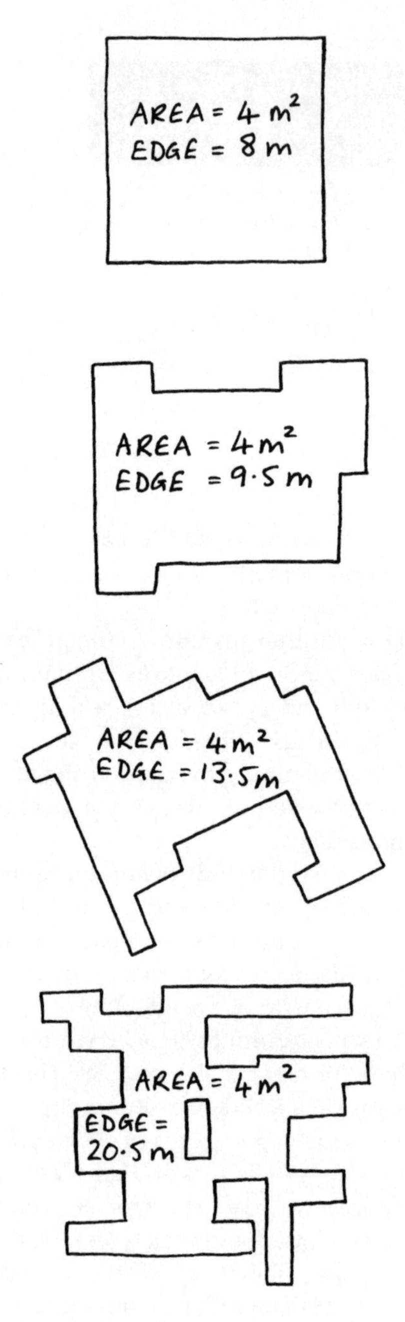

Getting the most edge out of one size of pond.

or at the centre of the meadow the number of species which can live well is more limited. So a badger, for instance, is much more likely to make its sett at the edge of a woodland for maximum feeding chances, and the safety of different kinds of cover. Many more plants will thrive in your garden if there's more edge.

Making as much edge as possible helps because it makes more use of the same area. No space lost, yet more output achieved. This is nowhere more important than in the limited space of a garden.

There is something very dead about a garden that is all straight lines. The flowing interactions of gardens designed in curves and hollows invite many more opportunities for variation in light and shade, wind resistance, privacy and mood, and consequently for greater variety of yield, be that food for the body or the soul.

Vertical Space ~

The plan and the map are marvellous tools to understanding space, but they are limited. By introducing symbols for contour we can start to represent three-dimensional space, but this is still a long way from the full possibilities offered by working with the land.

Drawing on paper is a good place to start planning the garden. For one thing it's a very cheap place to make mistakes! Designing in a sand pit, or with clay or Plasticine can bring us closer to visualizing the reality of land-

Starting with an 'all-square' garden design, a more adventurous and high-yielding plot is evolved.

scape. For the computer articulate there are now software programs for garden design, or maybe you can even write your own? As far as plants are concerned we should remember that all our crops operate in three dimensions, not two. The invisible part below ground is as important as the visible upper part.

It helps to see that plants function on many different levels. Roots are important anchorage and feeding mechanisms for nearly all plants. With some crops, such as parsnips or radishes it is the part underground which we largely see as the yield. Then there are those such as the *Sedums* (stonecrop) or *Veronicas* (Speedwell) which are largely at ground level, covering the soil

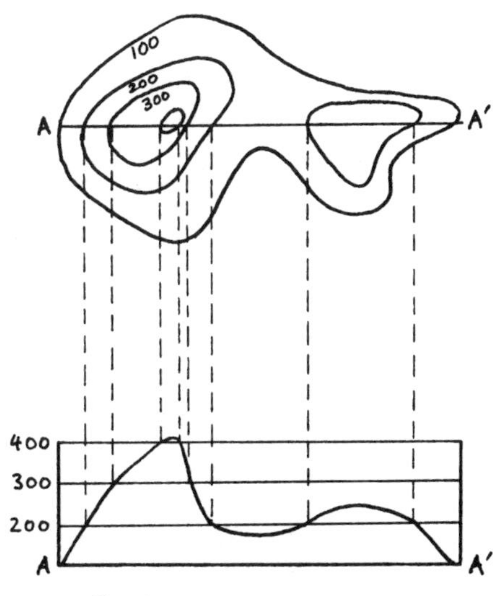

Contours on a map are a translation of three-dimensional space fitted onto paper.

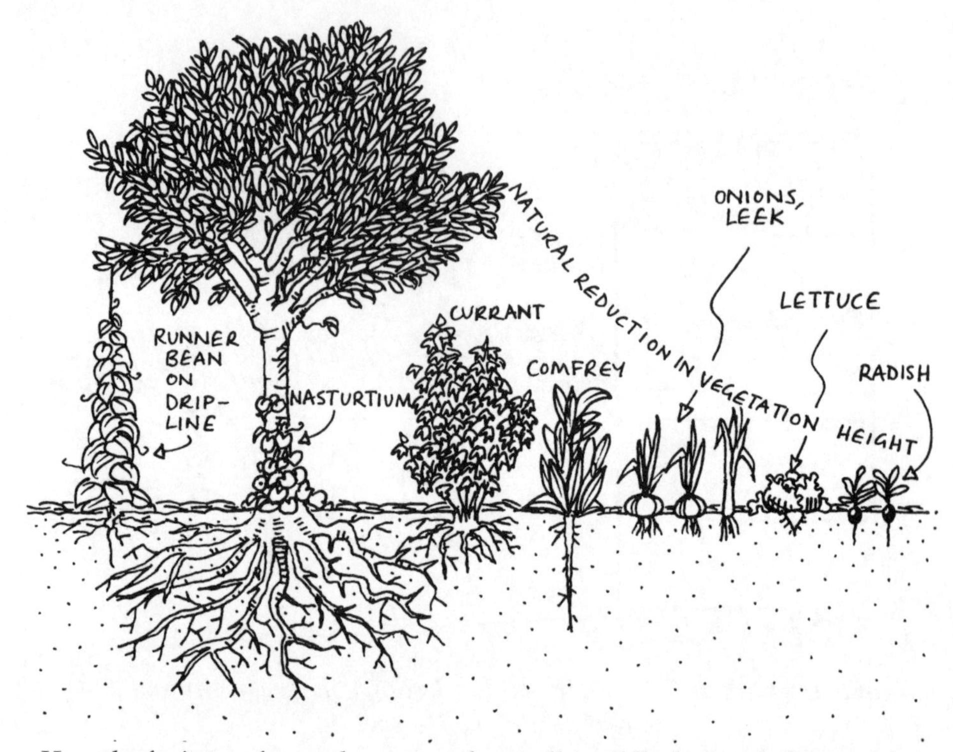

Here the fruit tree forms the centre of a small 'guild' of related plants, using every vertical layer of the garden as space for yield. The careful choice of species makes the plants mutually beneficial.

closely. Above this are the low herbaceous plants, such as mints or lettuces. These in turn are shaded by shrub layers, such as currants or *Philadelphus*. Higher still we come to the small trees, which in wilderness form the pioneers or edge species of the woodland. Here we find elder, hazel and so on. The tallest trees of the forest such as oak, pine and ash form the climax vegetation of the landscape.

We can consciously design a garden to step up from ground level to the tallest tree in stages which allow light and nutrients to be well shared by all the plants. We can use tall trees as frameworks for climbers, and as raised growing areas for mosses, ferns, fungi and all the other plants which like to get up off the ground.

As a classic example of vertical space as an opportunity for increasing yield, consider the surface area of the leaves of one mature deciduous tree to be around one and a half hectares. You can have three hundred such trees to the hectare of land, giving over four hundred hectares of green surface to the hectare of land. Suddenly the earth is no longer flat.

Bold use of the vertical can turn even a tiny garden into an enchanting space.

Expanding Upwards ~

Each of these ideas for creating or using vertical space has its own virtues:

- trees
- a trellis
- sheds
- fences or walls
- glass houses and polytunnels
- the house
- poles (e.g. telegraph poles, washing posts etc.)
- hedges
- down pipes.

They are useful because they have one or more of the following virtues:

- a living thing in itself
- leaffall providing food for garden
- yield fruit/nuts
- provide shade
- supports seats/tree house
- supports climbers
- place for animal life (e.g. bird or bat boxes)
- trap sunlight and warmth
- provide micro-climate (e.g. drier/wetter spot)
- create privacy
- create shelter from wind
- make vertical space for planting off ground.

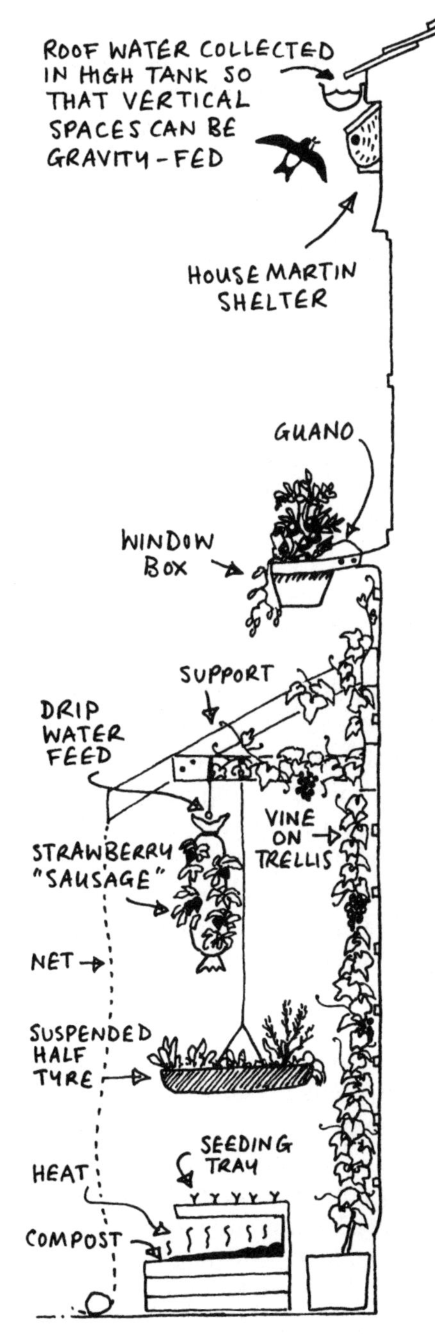

Creative vertical space usage.

So, for instance, if we're going to plant a hedge, don't just go for that quick growing *Thuja* or *Chamaecyparis*. Sure, they'll grow fast and within a couple of years you can celebrate the witch's sabbat on the back lawn (if you want) without your neighbours spotting a thing. *But*, they're very hungry plants and nothing will grow next to them. They'll shade out half your garden and will carry on growing like billyo even when they've got to two metres where you wanted them to stop (both trees and shrubs should be planted with an eye to their *mature* height). And they don't *yield* much. So why not make your hedge of cordoned fruit trees and get blossom and apples as well?

Look very carefully at every vertical opportunity in the garden. Few of us have such large domains that we wouldn't welcome the opportunity to fit a little more in.

Time as a Dimension

Many gardens sit out the winter as bare dug ground. Brown earth for four months of the year is neither aesthetically pleasing nor productive. Indeed it is counter-productive, for throughout the period of bare fallow the soil life is being starved.

There is really no such thing as the 'growing season' unless it is 365 days long. Plants have living processes at work all year, even when you can't see them 'growing'. The secret of successful growing is to plan the garden to give successions of yield throughout the seasons. Ground should be used throughout the year: bulbs can be followed by summer crops, which can give way to winter green manures. An area that is shady in summer may have plenty of light in winter and spring. You don't always have to wait till one crop is cleared before planting ('under-sowing') the next. It is also necessary to plan ahead across the years, so that as trees and shrubs gain in height and spread, the under-storey gives way. Think of ways to fill the garden with produce and sensuous delights all year round. For example:

Autumn

- trees for nuts and fruit (apples, pears, walnut, hazel, late raspberries)
- trees and shrubs for colourful foliage (maples, *Ginko*, *Prunus* spp, *Sorbus* spp, *Liquidamber.*)
- late flowering bulbs (autumn crocus, cyclamen)
- vegetables (cardoon, autumn cabbage, beetroot)

This is also the time to plant over-wintering crops including winter field beans, and green manures like grazing rye, tares and a final cover crop of mustard.

Winter

- winter foliage plants
- hardy salads (chervil, greens-in-the-snow, corn salad, winter radish, spinach)
- *Brassicas*
- trees and shrubs with attractive winter bark (*Cornus*, *Prunus* spp, birches) or that keep their fruit through the winter (*Cotoneaster*, *Sorbus*)
- very early bulbs and winter flowers (snowdrops, hellebores)

February is the leanest month of the year for a Northern hemisphere temperate garden, so designing for a good garden at that time of the year is a special challenge.

Table 4 Winter Salads

Salads can be grown all year round. This selection will provide good winter food.

Common name	Botanical name	
Japanese onion	*Allium cepa*	A
Welsh onion	*Allium fistulosum*	P
Leeks	*Allium porrum*	A
Garlic	*Allium sativum*	A
Angelica	*Angelica archangelica*	B In second year
Chervil	*Anthriscus cerefolium*	A
Celery	*Apium graveolens var dulce*	A
Land cress	*Barbarea praecox*	P
Beetroot	*Beta vulgaris*	A
Swiss chard	*Beta vulgaris var flavescens*	A
Mustard	*Brassica alba*	A
Pak choi	*Brassica campestris var chinensis*	A
Pe-tsai	*Brassica campestris var pekinensis*	A
Mustard spinach	*Brassica campestris var perviridis*	A
Kale	*Brassica oleracea convar acephala var sabellica*	A
Calabrese	*Brassica oleracea convar botrytis var italica*	A
White cabbage	*Brassica oleracea convar capitata var alba*	A
Red cabbage	*Brassica oleracea convar capitata var rubra*	A
Mizuna	*Brassica japonica*	A
Marigold	*Calendula officinalis*	A
Endive	*Cichorium endivia*	A
Chicory	*Cichorium intybus var foliosum*	A
Winter purslane	*Claytonia perfoliata*	A
Salad rocket	*Eruca sativa*	P
Lovage	*Levisticum officinale*	P
Purslane	*Portulaca oleracea*	A
Winter radish	*Raphanus sativus*	A
Sorrel	*Rumex acetosa*	P
Salad burnet	*Sanguisorba minor*	P
Dandelion	*Taraxacum officinale*	P
New Zealand spinach	*Tetragonia tetragonioides*	P
Winter pansy	*Viola* spp	P

Spring

- bulbs (crocus, bluebell, wild garlic)
- early blossoms (*Forsythia*, lungwort, *Anemone*)
- early salads (seakale, Witloof chicory)

Suddenly the garden shoots up and gets away before you know it, making:

Summer

A riot of growth. So it's good to design a summer garden which is relatively maintenance free, with plenty of perennials, self-seeding annuals, continuous ground cover to reduce weeding and evaporation (and therefore watering). If you're limited in space try growing things that are more expensive or hard to get from the shops, which you like, rather than just the 'easy' options.

It's also important that we are conscious of changes through the day. So, for instance, some plants profit from early sun (day lilies, sunflowers), others will need to be protected from early sun. Pear blossom is out at frost time and an early blast of sunshine can kill it, whereas a slow thaw through the day will leave it unharmed. We need to leave resting points in the garden where we can be shaded from the full heat of the midday sun. That's particularly true of play areas for toddlers.

Appropriate placement means that we're working with nature rather than against. For instance, if the frost lies later in a certain corner then put hardy plants there, or leave a hole in the adjacent wall or fence to help the frost run away downhill. It may be that knowledge of these details will improve over succeeding seasons, so a garden plan needs to be flexible enough to accommodate improving understanding of the specific site.

Table 5 Some Plants that Attract Beneficial Insects

Plants that attract beneficial insects.
(Plants must be allowed to flower and seed for maximum benefit)

Common name	Latin name	Type	Attracts
Garlic mustard	*Alliaria petiolara*	A	Orange tip butterflies
Borage	*Botago officinalis*	A	Bees
Butterfly bush	*Buddleia davidii*	P	Butterflies
Buckwheat	*Fagopyrum esculentum*	A	Bees/Hoverflies
Lemon balm	*Melissa officinalis*	P	Bees
Devil's bit scabious	*Succisa pratensis*	P	Bees
New Zealand spinach	*Tetragonia tetragonioides*	P	Hoverflies
Umbel family	*Umbelliferae*	A/B/P	Hoverflies
Nettles	*Urtica dioica*	P	Caterpillar food

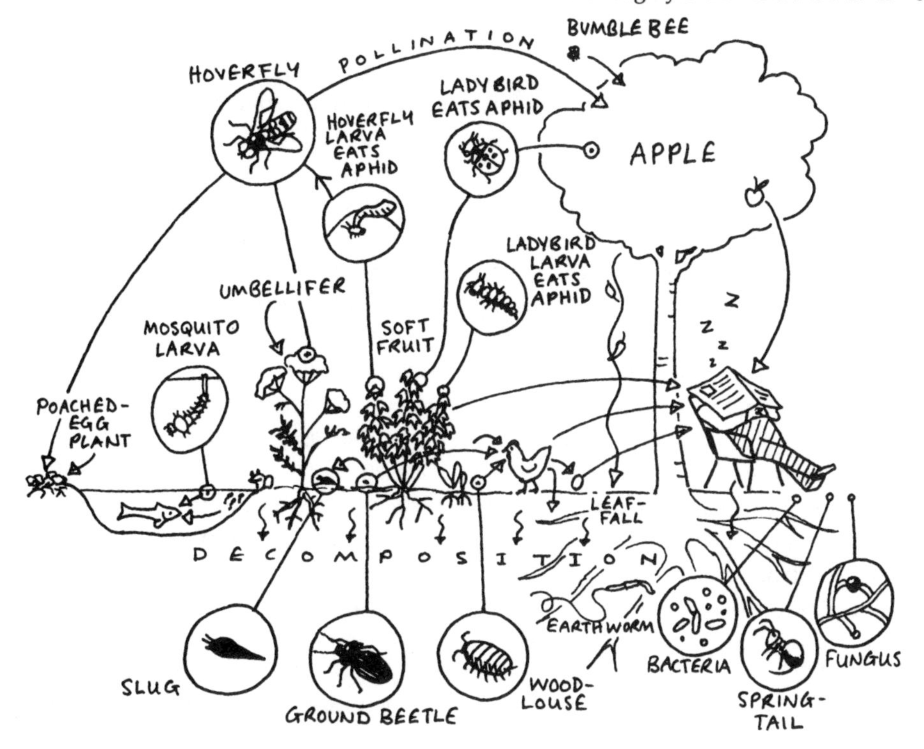

The web of life in a garden.

Everything is Connected

These kind of judgements in controlled management of garden space are more successful when we are more aware of the way that all the elements of our garden work together.

This area of knowledge has had a lot of currency in recent years as 'companion planting'. So we learn, for instance, that wormwood (*Artemisia* spp), deters cabbage white butterflies, and is therefore a protector for *Brassicas*. In fact it also exudes a growth suppressant which limits the yield of adjacent cabbages.

So watch out for the reliability of such folk wisdom, and temper it with knowledge based on your own experience.

Reading that buckwheat was a good attractant for hoverflies, which are voracious devourers of *Aphis* (Greenfly), I thought to control the pest population in my garden. I soon noticed that spinach and *Umbelliferae* flowers worked better.

The umbel family (which includes angelica, lovage and parsnip amongst its useful members) has tall sprays of flowers, and I am led to conclude that any small tall flowers are attractive to the industrious hoverfly, and so balance the insect population of the garden. Now I am happy to let a proportion of vegetables in the garden go to seed, and discover the beauty of many of these plants at the end of their season.

Relationships and connections can be seen in time, space and between the plant, insect and animal worlds. It sounds obvious to say that no living organism functions in isolation, yet many gardeners ignore the value of placing elements of the garden where their life processes will be mutually beneficial. Chickens and ducks are great weeders and pest controllers. Why spend time and energy on bean poles, if the bean can be trained up a living tree? Fruit trees cannot bear fruit without pollinators–though plenty of garden centres sell fruit trees without explaining this. And so on.

As a guideline, it should be possible to think of five uses for everything in its place in the garden. Beginners can practice with three. An easy example is a hedge: it can provide fruit, windbreak, firewood, shade, make a sun trap, be a wildlife haven, keep stock (dogs, people) in or out, or even give privacy. Applying this principle throughout the garden makes a web of life that is extremely strong and high yielding.

Working with Nature

Suburbs can offer a biologically rich blending of exotic and native plants and animals, a diversity that ensures the ongoing ecological vitality of the suburbs and the delight of its inhabitants.

ROBERT KOURIK, *EDIBLE LANDSCAPING*, 1986

So the sustainable gardener will aim to build systems which work because they are designed by observing nature. That means going with the flow. 'Work with what you have' would be a good first principle. Look at your garden and decide what it's good for now. A new house with a rubble-filled plot suggests an instant rockery. A neglected garden full of weeds has plenty to tell us about the condition of the ground–don't make your first response to dig the lot in.

Certain plants favour certain soils, so rather than change the soil it may be better to fit the plants to the existing environment. If your soil is of an extreme type (e.g. very acid), favour plants suited to the conditions (so bilberries will love a very acid plot).

Place plants where they'll be happiest: put shade-loving plants, for example, in shady corners, and sun lovers where it's sunny.

Of course this means you are limited in what will grow well, but that's inevitable. It has the advantage that you get a good garden for minimal effort.

Liabilities into Assets

Whether something is a problem or an opportunity says more about our way of thinking than it does about the situation itself. To put it another way, every liability is an asset if looked at from a different point of view. Most junk can be put to good use in a garden if we use our imagination. Here are some examples.

Potato Towers
Collect five old car tyres of equal size. Stack two above each other and fill up with kitchen compost. Cover with a thin layer of soil, grass clippings or straw. Plant five potatoes on this. Potatoes are the swollen stems of the 'haulm' or green part of the plant. So as the potatoes sprout and grow keep adding tyres, and filling up with soil or organic matter. At harvest time simply lift off the tyres and pull out the spuds. The contents of the tower will be rich black soil. This works, even on a concrete yard. Six towers in a hexagon, with the inside painted white to reflect the light, make an excellent heat trap for growing tomatoes in the centre.

Pallets
Five pallets tied together in a cube without a lid make the quickest and neatest compost heap. If there's room in your back yard add another four alongside to make a double cube for rotating the waste store. When one heap is full plant a marrow on top, and keep it watered. A piece of polythene left in a corner of the yard will accumulate brandling worms (small, thin, very red worms) underneath. Put them in the heap to speed up digestion of the contents.

Barrels
Use as rain butts, or for contained worm compost digesters. Mounted horizontally on a pivot they make excellent rotating compost bins for smell-free digestion, and a quick harvest. Barrels make good play equipment, ponds, drums, strawberry towers (with holes cut in the side), mushroom beds or chicken sheds. Just make sure that the previous contents were benign for garden usage (i.e. no poisons).

When confronted with a frustrating problem in the garden, therefore, just take five minutes to think of why it's also an asset.

Counting the Harvest ~

In the sustainable land use which is our only option for the future, we have to count *true* yield. Yield is the sum of what you get out of a system, *less* what you put in. Forty kilos of potatoes from an afternoon's work and no chemical inputs is a higher yield than two hundred kilos from the same area achieved by double digging, weeding, watering and adding chemical nitrates.

Chemical growers tend to run down organic systems as being of 'low yield'. They never account for the damage they do to the land as part of the cost of their outputs. Nor do they weigh the health burden of all the chemicals we ingest. Soaring asthma rates in Britain can be directly connected, not just to atmospheric pollution, but to pollution of the food we eat.

Mono-cropping leads to inflated results as well, because the yield is only counted for this season. Multi-cropping is much higher yielding in the long run, although fertility in the system may take years to build up to its peak. But when it does so, it stays there, and the crops and the soil are healthier.

Diversity of output leads to less disease risk, and greater availability of crop if one type fails.

Summary ~

The physical patterns of nature are worth emulating in the garden because they trap energy. Inventive use of three-dimensional space, time niches, and good inter-relationships are what we are looking for here. Old thinking can blind us to easy productive solutions, but creative patterns of thinking can be learned and replayed for human benefit–the gardener's not least of all.

three

What can I do in a day?

The reward of the thing well done, is to have done it.

RALPH WALDO EMERSON (1803-82)

Early success encourages any gardener. These suggestions are for one-day projects which show results quickly. You don't need to be an 'expert' to succeed.

There's no substitute for considered action. Doing something is a much better way of acquiring understanding than reading text books. All the following one-day projects are possible with very limited tools or knowledge. For most a spade, a fork, a trowel and a wheelbarrow are all the tools you need, although there's little you can't do with your bare hands if you want. Yet each, with a little observation, teaches new lessons to the inexperienced gardener, or sharpens the wits of the knowledgeable. However, a little time spent planning each venture, and getting the materials together will make them more successful.

All of these ideas can be more productive and more fun if you carry them out with someone else. Two to five people is a good number. You can set up a neighbourhood one-day garden 'hit squad' and take turns to enhance each other's back yards. Don't try and do too much at once. It doesn't matter if you don't do the whole garden at one go. It's more important to target a small area, and have the satisfaction of completing it.

Containers on Concrete

Gardens can be made anywhere. I met a woman once who told me she made a lawn at her temporary home in the Kalahari Desert by getting up at dawn and searching out individual blades of grass with a pair of tweezers. You may be able to think of more useful occupations. The point is that even the most trying circumstances are a challenge which can be overcome. You can make an attractive and productive garden in one day–even if your back yard has no soil.

For the purposes of the exercise we'll assume we're looking at a concreted back yard that's enclosed by two-metre-high walls in a town. It measures the width of a small house (say eight metres) by a similar length. The access is by a back door from the kitchen. The size is typical of lots of urban house designs, although often the yard is actually 'L' shaped.

Either way there's still plenty of opportunity for gardening here. Firstly you'll need containers to stand on the concrete, and secondly you'll need organic and mineral material to fill them with.

Urban situations are rich in 'junk', so this method works for the person with no money at all. Those with the resources can invest in any of the delightful and attractive range of purpose-made fancy tubs on the market–from barrels to ceramics. For the less well heeled (or simply thrifty) the containers can be recycled wood, plastic, stone or ceramic. In some cases wire or glass will do, but make sure these are set up so no-one's going to get cut by sharp edges. See that they are as unpolluted as possible: a good rinse out is all that many containers will need, but containers for chemicals, petrol, paint and so on, are probably unsuitable. Here are some ideas:

- Old building materials are great: chimney pots, old 4 inch salt-glazed drain pipe, plumbing bends, 'Belfast' or 'Butler' sinks, and lavatory pans.
- Barrels (check out local factories involved in food processing for edible grade polythene tubs).
- Heavy timber, such as railway sleepers, that can be stacked to hold deep beds.
- Old bricks (the more decorative the better).
- Pallets.
- Car tyres.
- Rotted rounds of tree trunk with hollow centres.
- Wire baskets for hanging.
- Heavy duty glass carboys (even better indoors)–but make sure they're clean inside.
- Wickerwork made by weaving sticks cut from an overgrown shrubbery.

You can collect and prepare these over a period of time, or you can make a one-day raid on all the skips in the neighbourhood. Containers covering only two square metres can produce salad and herbs for a couple of people right through the summer, so you don't need to cover the yard at the first attempt.

To fill the containers you need soil. This is a very good lesson in understanding the composition of good soil–because your garden is going to make its own! Soils vary enormously from one place to another, but they're always composed of a mixture of sand, clay, silt and organic material in different proportions. Within that mixture a living soil also has a huge array of living organisms–from microscopic bacteria to earthworms. A good organic soil has perhaps ten tonnes of living creatures to the hectare.

Firstly create good drainage by layering the base of each container with 3cm of chippings. Old rubble or broken crockery is fine. If there is any soil in your yard, heap it into one of the containers. Also start to save and keep any organic waste from the house. That means all kitchen scraps: vegetable peelings are ideal, animal bones will take a lot longer to rot. A small proportion of waste paper can be used, but avoid coloured dyes, especially red and yellow, which contain heavy metals. Newspaper is alright because the black ink is carbon, not lead. Cotton and wool waste is useful, so are hair and nail clippings.

If your own household supply is meagre then supplement it with leavings from your local hairdresser or fruit and vegetable shop. Riding stables in towns are usually glad to get rid of unwanted manure. Pet shops may be in a similar position; straw and sawdust will help.

Many households just throw away their compost for the garbage collection, so help yourself to lawn mowings and hedge clippings on collection day, or make an arrangement with friendly neighbours to take away their unwanted organic trash. Many local authorities don't compost park refuse, preferring to burn it. They'll usually be glad for you to take away raked up leaves and save them a job. Obviously good relations are maintained by asking a responsible individual first!

Out of the concrete comes forth greenness.

Sand, old moss from roofs, lime plaster (*use only a little*) can all be found where building work is going on, as can the odd unwanted bucket of soil dug from foundations. These will add to the mineral content.

The best fellows you can import to finish the job off for you are earthworms. You can buy starter packs for making worm composting systems, or you can simply collect worms wherever you find them. I've yet to clean a gutter out without finding a few pink wrigglers. A walk under the trees with a plastic bag in your pocket is an easy way of starting your own worm fraternity. Don't worry what people *think*, worms are very hygienic. You also get used to picking them up pretty quickly when you realize what good they do.

Pop the worms on top of your containers as they fill up with material–they'll soon burrow in. They digest all that organic material and turn it into topsoil.

When the containers are full, you can think about planting. A good starter crop is potatoes because they rapidly turn the rough compost into black friable soil. (See the potato tower in Chapter 2 for details.) Large seeds will plant better in rough-stuff than fine ones, so beans, peas, onion sets and spinach are good starters. Otherwise buy plants which are pot or cell grown. A dozen cabbages can usually be bought in a strip of cells from spring through to autumn time. Their own root ball is intact, so they'll establish quickly in your compost tubs.

Some spreading plants, like nasturtiums or periwinkle, will bring colour and coverage to your grey backyard. If you're just starting up, then get into conversation with other gardeners in your neighbourhood. I've yet to meet a gardener that would begrudge another a cutting or a bit of root when they're splitting up an over-full bed. I've also yet to meet a gardener that didn't want to give someone less experienced a bit of advice!

Most seed packets will tell you exactly how to plant the contents. I divide the quoted yield by a handsome figure. If every seed was as 'prolific' as they seem to suggest I'm sure all suburbia would have reverted to jungle by now. I also find that *most* plants are a little more tolerant than they suggest. However, packet instructions are trying to direct you to the best possible result. Experiment with the bit you've got spare rather than the only packet of seed you could afford this week.

Table 6 Living Mulch Plants

Quick ground cover plants form a living mulch.

Yarrow	*Achillea millefolium*	P
Bugle	*Ajuga reptans*	P
Lady's mantle	*Alchemilla vulgaris*	P
Thrift	*Armeria splendens*	P
Marigold	*Calendula officinalis*	A
Wild strawberry	*Fragaria vesca*	P
Stonecrop	*Sedum* spp	P
Nasturtium	*Trapaeolens major*	A
Crimson clover	*Trifolium incarnatum*	A
White clover	*Trifolium repens*	P
Periwinkle	*Vinca* spp	P

To gain height in the yard fix poles, wire or string up the walls and train climbers, such as runner beans on them. Hanging baskets are another way of achieving height–but remember you have to water them. All of your garden will need an occasional sprinkle, and lining with material such as old felt carpet underlay will help to keep moisture in. If the water can be collected from the roof so much the better.

For people worried about urban pollution, remember that green leaf food is best, as roots tend to concentrate the nasties. A good rinse in clean water will remove air-deposited pollution.

For maintenance, just keep topping up the mulch in the tubs, harvest the goodies, and replant when necessary.

The Tree Garden (1)

There are lots of neglected old gardens around the world. The new owner moves in and finds a sea of grass and weeds, and an old fruit tree someone planted fifteen years ago, and then forgot about. 'When are you going to clear the garden up?' s/he says to him/her. Lock away despair, here is an easy solution. The one-day tree garden.

The first rule in any garden is 'start small'. So rather than tame the entire wilderness, just concentrate on that one centre of interest, the tree. If it's overgrown it may need cutting back. If it needs a heavy prune wait until the autumn for that. Take the tree back to four main branches, and thin out the canopy so that light can get all around the fruiting buds. If major surgery can be done before you make the garden, then you're less likely to mess it up later.

The idea now is to create a circular garden around the tree. It will use the tree as an interceptor and provider of useful nutrients, and build a community, or 'guild' where each of the member plants is placed for maximum benefit to its neighbours.

Water concentrates in two areas: at the base of the tree's trunk, and at its 'drip line'. This is where rain drips from the edge of the canopy. Under the 'umbrella' of the tree is relatively dry.

To get best benefit from this self watering opportunity dig a trench half a metre (18in) beyond the drip line to a depth of 25cm (10in). Keep any loose soil in a heap for later. Any weeds or turf can be used to build a slight lip on the edge of the bed (see illustration). As the tree grows the bed can be extended to stay ahead of this line.

Cover the surface of the bed with a double layer of cardboard, and then bury this completely in 10cm (4in) of nitrogen-rich material, such as compost or well-rotted manure. Add to this any loose soil remaining from the trench. Add a top layer of carbon-rich material such as fine twigs, wood chip, straw or plant cuttings. This is known as a sheet mulch system.

The end result can be fairly rough, for again, within a season the surface trash will rot and be digested to fine crumbly soil. This is more fully explained in 'Grass into Vegetables', p. **41**. Once the bed is made *never* walk on it if you can

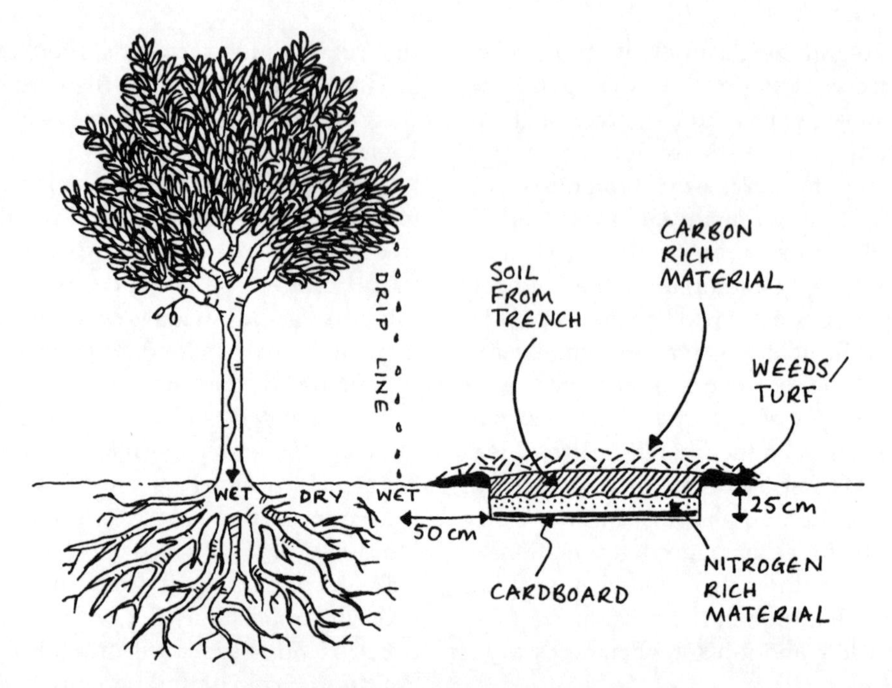

The tree garden is built to trap as much energy as possible, and to offer best mutual resistance to disaster. It needs little work to create and maintain it.

avoid it. Make pathways for access or you can place stepping stones or rounds of timber for walking on.

Next, plant up with appropriate species to create a cross section which, like the illustration (p. **26**) uses all the layers of the vertical growing space. Plants should also be chosen for range of yield and mutual benefit. In the example shown the following yields are intentional:

1. The tree harvests rainwater, and gives leaffall which is rich in nutrients mined by the tree's root system. It also gives support to climbing plants.
2. The beans use the vertical space of the tree, and being leguminous, and therefore nitrogen-fixing, help feed the other plants.
3. Marigolds are not just sunny flowers. Their root exudations deter harmful nematodes (tiny worm-like creatures) from attacking other plants. Both flowers and leaves provide salad produce for a large part of the year, rich in vitamin E.
4. Onions also provide a healthy root zone. Chives are of the onion family too, and offer a 'cut-and-come-again' alternative.
5. Land cress is perennial, and by seeding

itself builds continuous ground cover giving winter salad.

6. Chervil is another winter salad, and being part of the umbel plant family it has tall umbrella-like flowers and is a good attractant of beneficial insects.
7. The mulch helps feed the tree, and being resistant to drought helps the tree resist mildew in dry summers, whilst avoiding the need to water.

This is a pleasant day's work, and the bed can be extended later to become an oval encompassing a second tree, or can become part of the process for making an old orchard into a forest garden–but more of that later...

If you don't have an old tree to tend to, a young tree can be planted as the central interest of a bed started from scratch.

The Tree Garden (2) ~

An alternative to the above is inspired by the sight of millions of Christmas trees heading for the town dump every January.

Buy a rooted Christmas tree, and after the Yuletide festivities make a bed centring on the tree as above. The bed will be best for having acid soil, which can be encouraged by a very high proportion of organic matter in the mulch, and preferring sand for any soil used. Add the clipped up branches of all your neighbours' (dead) Christmas trees as mulch.

Many standard vegetables will do well in a mildly acid bed. Other suitable plantings might include:

- Shrubs: blueberries, bilberries, indigo, heather
- Herbaceous: primulas, thymes
- Ground cover: Alpine strawberries, stonecrop
- Bulbs: wild or cultivated garlic, snow-drop, fritillary
- Climbers: honeysuckle, *Trapaeolum speciosum*

Grass into Vegetables ~

Lawns are great places for sleeping in the sun, playing with the kids, or (mostly) consuming huge amounts of energy in creating something perfectly green, evenly short, and practically useless. In other words, they're something which we can easily turn to better use. Here is a one-day garden which can be made with unwanted lawn.

Growing areas can be made quickly by using sheet mulch techniques. We looked at this technique in making the tree gardens above **(p. 39)**. Thick cardboard, newspaper, and old clothes and carpets (but no synthetics, please!) are laid out. These suppress the weeds. This can be done even over long meadow grass, thistle, and docks. Over this layer put a hand's depth of material rich in nitrogen: compost, manure, kitchen scraps, old leaves. Then cover the whole with carbon-rich material: straw, clippings, wood chips and shredded paper. It may look rough at first, but soon degrades to give a fertile looking soil.

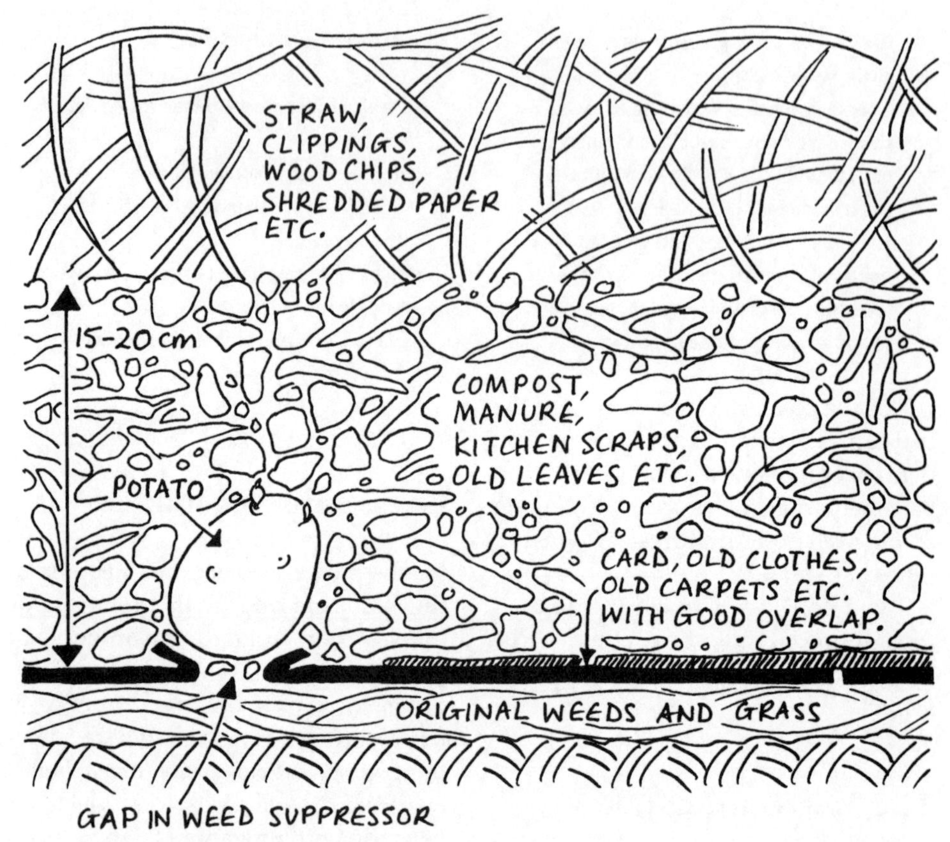

Section through a sheet mulch showing the layers.

Sheet mulching is best done in the autumn, leaving time for establishment for spring plantings. When frosts are over, plant the mulch up with seeds that can stand the rough effect–such as potatoes, beans and peas, *Calendula*, buckwheat or onions–by pressing the seeds down into the nitrogen layer. Plants grown from finer seeds (e.g. *Brassicas*) can be started out in containers and planted up when fist sized.

A cross cut through the underlying sheet made (carefully) with a stout bladed knife will enable deep-rooting species to get a tap root down quickly. The rough mulch can be pulled away to make space for them, or they can be set into small pockets of topsoil in the straw. Yields may not be too high the first year, but after one season a deep humic layer of topsoil is created which is full of earthworms, and future yields increase.

When judging yield remember that this method completely eliminates the backbreaking work of digging and weeding a previously untamed plot. Maintenance can be as simple as continual mulching, although you can weed out any undesirables that break through the sheet part of the mulch. I have made many sheet mulch gardens in an afternoon on many different sites, and the speed of bed creation is such that even the novice gardener can be up and running by sunset of the first day.

Some people complain about subsequent problems with slugs in the mulch. In general, an over-abundance of any pest indicates that something in the system is out of balance. Slugs like acid, moist conditions, and hate sulphur. Wood ash or pure coal soot will deter them. Occasional annual liming will prevent beds from becoming too acid. Aromatic and spiky plants are thought to deter slugs, so frequent interplantings of herbs will help to keep the garden healthy. Old yoghurt pots buried to the brim and filled with stale beer give the slugs a happy end. Hedgehogs, wild birds, hens and ducks like to ride on the back of the slug population.

For subsequent maintenance keep the surface mulched as much as possible, and certainly out of the warm season. The mulch also suppresses weeds. Beginners tend to be timid in the depth of mulch they use. It all rots down pretty quickly, so I advise boldness. Unwanted mulch can be simply raked up and reused elsewhere in the garden.

Making the bed with limited edge at first is helpful. Shaped beds can always be made after the original sheet mulch has rotted down. As with the tree garden, never walk on the bed if you can avoid it. If you must walk on the surface, use a board to stand on which spreads your weight and avoids compacting the soil. Also never dig the bed if possible. The layers are carefully constructed, and inverting the soil destroys the balance of a self-managing system.

The German Mound ~

This is a clever way of using up unwanted woody cuttings, making a bed that will feed itself for a long time to come, and a way of adding variety to a flat garden.

You need a supply of woody material (hedge and fruit tree clippings are ideal) big enough to make a round pile a metre across and half a metre high. Stack this alongside the chosen site and then strip the turf and topsoil away from a metre circle, keeping the two separate.

Heap the sticks in the circle, standing on them to compress them. Cover with the inverted turf, and then the topsoil. Water well to help the heap settle. Now you can plant up. The mound is ideal for a tree or shrub as its centre, and this may be planted more easily into the pile of sticks whilst the mound is built. If doing so, make sure there is plenty of soil around its roots. Air is trapped in the twiggy pile and the soil will take a while to settle through these gaps.

You now have a round raised bed ready for planting. The centre of the bed will rot down very slowly, providing a long slow release of nitrogen. You've also prevented an unnecessary bonfire to burn all that useful stuff.

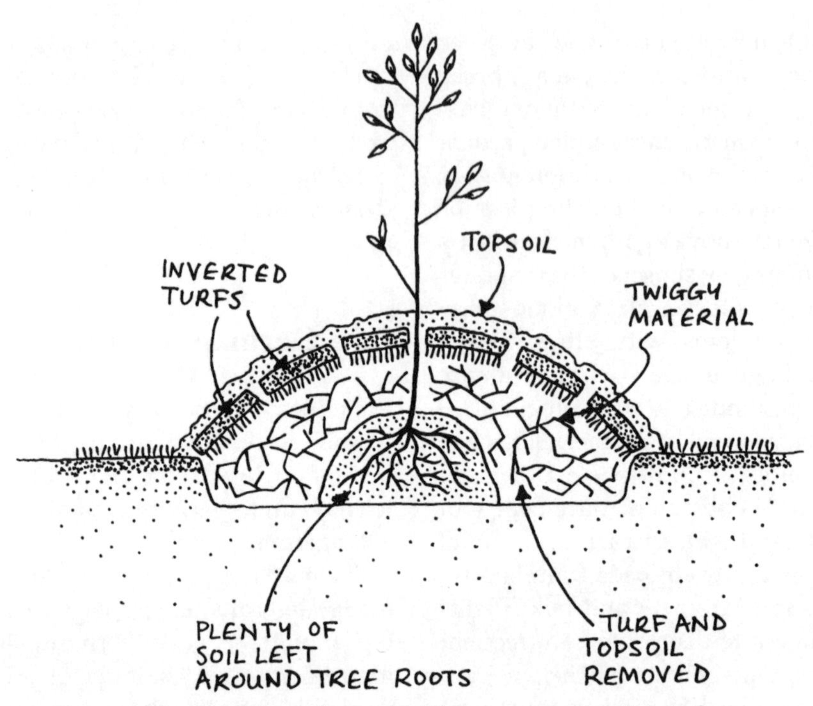

Unwanted garden rubbish becomes a useful and attractive growing space.

Children's Garden ~

With a children's garden you need to decide if it's a garden you manage for the children, one you share as a learning opportunity, or one where juvenile anarchy runs riot. First, it's good to think about what children do in a garden. Not many young folks spend hours weeding and planting a plot, unless you have the time to do it with them. If you have, it's the best way of fostering their interest.

Kids do like to spend lots of time burning up energy, and playing fantasy games. Put space for young children near the house, with at least partial shade. If there are bigger kids who want to charge about and throw balls, make sure fragile plants are protected (e.g. a protective trellis).

This garden is designed for two to six year olds, but you can adapt it to any client group. There's no reason why adults can't have playgrounds too! Whilst remembering children, therefore, don't forget adults: keep a sunny and a shady spot clear for sitting outdoors in appropriate weather. Surround it with colour and scent (or whatever your preference) to make it a pleasant place to be.

If you want to include more robust play furniture (see Chapter 6 on 'play space'). Some scrap materials suggest good uses as ready-made play aids. Old dinghies and short lengths of ladder are good examples. If you see such things around, grab them quick. Parts of this design work really well where there's a natural slope to the garden, which makes sliding and climbing easy.

In this design children are taught not to walk on growing areas, and have plenty of opportunity to make this instruction into a game. They can also practise with relatively easy plants.

Plants that kids like to grow are mostly quick germinating, easy to grow, and fun to pull up or pick: mallow, nasturtiums, marigolds, daisies, sweet peas, potatoes, carrots, mustard and cress, radish, lettuce, peas and beans, and strawberries are good examples.

One city farm has fitted out a see-saw so that it operates a water pump–an excellent idea for useful yield from play. You can have great fun thinking of multiple uses for structures in the

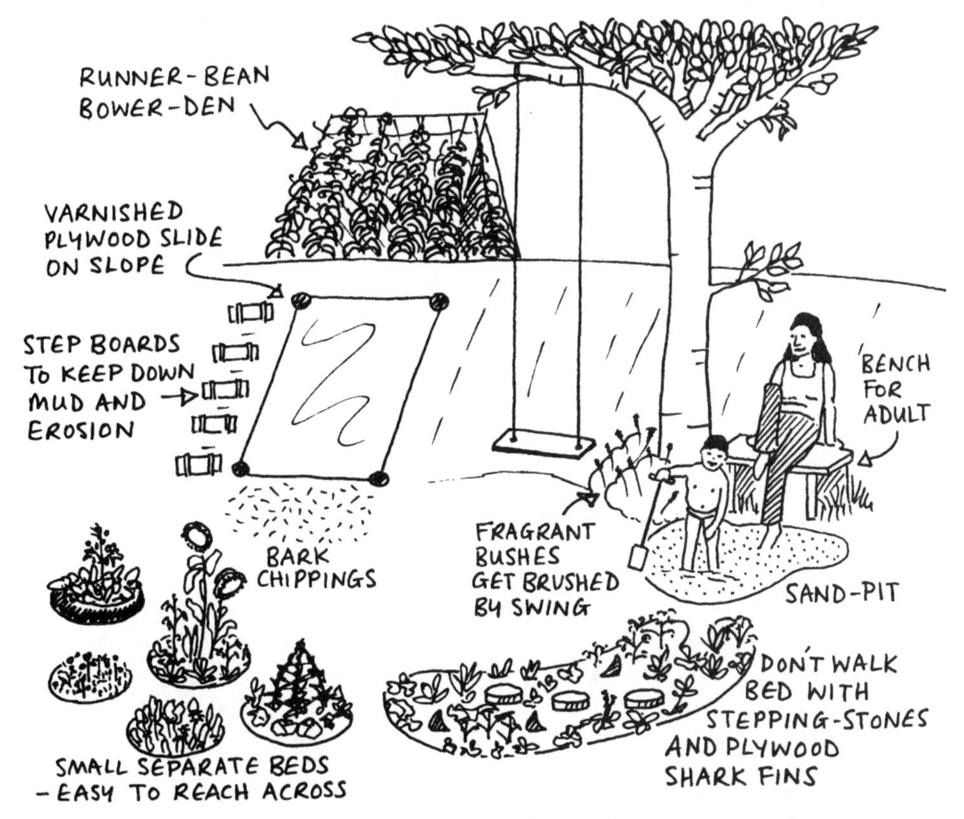

Working with children in the garden can be very rewarding. They'll like an introduction to gardening that's gentle. This garden tries to combine play with learning.

garden. In one garden where I designed a trellised walkway, the owners added in a swing from one of the horizontal cross pieces. Any upright can support climbing plants, which can in themselves be attractive: runner beans are a natural for children. They were first introduced from South America because of their attractive flowers. It was only later that people realized you could eat them.

Garage into Garden ~

Many gardens have a garage or wooden shed, which might be thought of as 'unsightly'. Sheds are, however, extremely useful as supports for the extended garden.

You can:

- Add on a greenhouse, using the garage wall as a suntrap, and gaining heat for when you're working in there on cold days.
- Turn flat or gently sloping roofs into gardens themselves. Cover with a waterproof membrane, fit edging boards, earth up and plant up.
- Collect rainwater from the roof for your own water garden, or simply for irrigation.
- Trellis and grow plants up the side of the shed.
- Notice which side is shady and which side is a suntrap, then plant the adjacent area with plants that like those conditions.
- Hang baskets from the eaves.

The only limit on productivity is our imagination.

Table 7 Useful Climbers

Useful climbers.

Acacia	*Acacia dealbata*	TP Nitrogen fixer
Siberian Kiwi fruit	*Actinidia arguta x meader*	P A hardy fruiting variety
Kiwi fruit	*Actinidia chinensis*	TP Male and female plants needed to set fruit
Ornamental quince	*Chaenomeles japonica*	P Needs training
Travellers joy	*Clematis vitalba*	P Medicinal
Convulvus	*Convulvulus cneorum*	P Attracts hoverflies
Moroccan broom	*Cytisus banderii*	TP Nitrogen fixer
Fig	*Ficus carica*	TP
Ivy	*Hedera helix*	P North hardy/bee fodder
Honeysuckle	*Lonicera* spp	P Woodland plant/shade tolerant
Hop	*Humulus lupulus*	P Bedtime teas/beer
Hydrangea	*Hydrangea petiolaris*	P North hardy
Winter jasmine	*Jasminum nudiflorum*	P North hardy
Myrtle	*Myrtus communis*	TP
Passion fruit	*Passiflora caerulea*	TP
Russian vine	*Polygonum baldschuanicum*	P Green manure
Robinia	*Robinia hispida*	TP Nitrogen fixer
Rose	*Rosa* spp	P Especially R.rubiginosa, R. canina
Dewberry	*Rubus caesius*	P Autumn fruit
Blackberry	*Rubus fruticosus*	P Many hybrids e.g. tayberry
Nasturtium	*Trapaeolus major*	A Edible
Grape vine	*Vitis vinifera*	P Select hardy varieties e.g. Brant
Wisteria	*Wisteria chinensis*	P Nitrogen fixer

Grey Water Reed Bed ~

This is a one-day design to give water of adequate quality for irrigating the garden, whilst reducing wastage; a bit of self-help green action.

Western countries are incredibly wasteful of water. In the Third World aid workers set targets of around 40 litres of clean water per person per day. In Europe and the US our average daily usage is 400–600 litres per person. Droughts in recent years have convinced many in the developed nations that our water supply isn't as secure as we thought.

In the UK we waste massive amounts of energy by insisting that all our water be of the same quality, and by mixing all our waste water, from the toxic to the slightly soapy, into one big system. Grey water is the waste from sinks, showers and baths. In other words it's had nothing dirtier in it than you, your

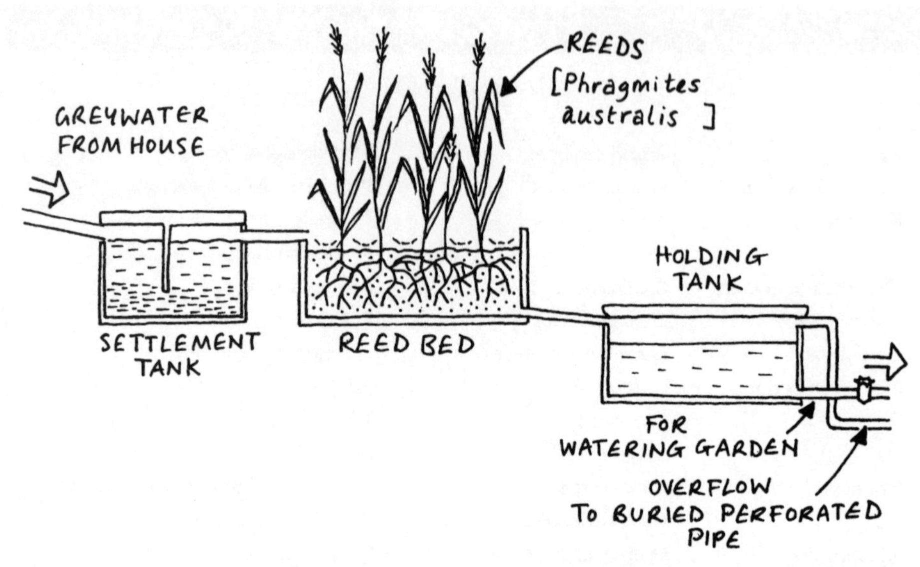

A grey water treatment filter/reed bed is a simple treatment system which offers irrigation-quality water from 'grey' waste.

clothes or your dishes. It shouldn't contain sewage.

By intercepting the waste pipe from your sink you can divert the flow into a compact cleansing system which will easily give water clean enough for the garden. The plumbing skills required are fairly basic. If you don't want to go that far, you can always make the system so that you empty buckets or bowls into it. This design is a miniature version of systems now being used to clean up municipal sewage and effluent from chemical works as big as ICI's plant on Teesside in the northeast of England, so don't worry, you're in good company.

Some helpful points. The higher the level of the water input, and the greater the fall in the garden below this point, then the more effectively gravity can be used to distribute the outflow. The reed bed itself simply needs to be planted with wetland plants native to the area you live in. Even cities have plenty of these growing in odd places. Reeds, rushes, irises and so on are ideal.

The outflow from the system can be directed to a pond, or to flood or drip irrigation. Flood irrigation means it simply overflows across a level area; trickle irrigation, that it is directed into pipes which water through perforations along their length. The outflow from the system should be free of all harmful bugs. Additional security can be had by never using the water on crops eaten raw (i.e. salads), but this precaution should not be necessary if only grey water input is used.

four
What have I got?

Never, no never did Nature say one thing and Wisdom another.

EDMUND BURKE (1729-97)

You can only start with what you've got. Thinking carefully about your natural resources is the key to good planning. Remember that every problem can be turned into an asset if looked at differently.

Gardeners love a challenge.

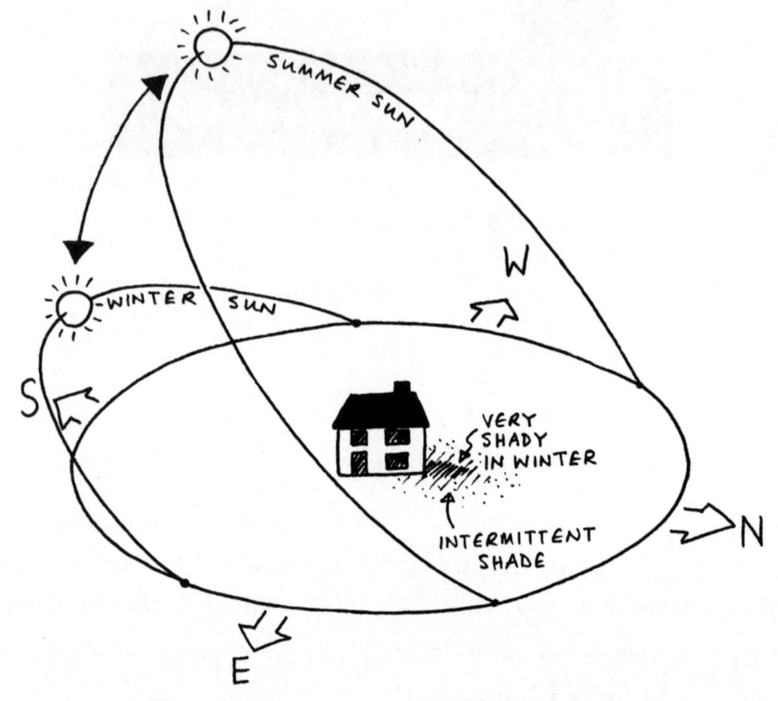

Varying sun sectors through the year.

Resources ~

Working with what you've got means regarding everything as a resource. When you look at the present situation as 'a problem', you're not only wishing your life away–you're turning away from a possible asset. So none of the physical aspects of your garden as you find it are inherently good or bad. They're all just how it is, and to avoid unnecessary work means accepting that at the outset.

Start by looking at the garden and just observing what resources it has and how they affect it.

Sun ~

Ask these questions:

- How much of the garden gets full or partial sunlight?
- How much is in full shade?
- How does this change through the year?
- If short of sunny spots, can anything be cut back to make more light?
- If the garden is all sunlight, does it need a shady spot? (Scientists agree the ozone layer above us is now holed for significant parts of the year; full sunlight is no longer recommended for the human skin, so shade is not just a matter of comfort if you want to enjoy sitting in the garden, it's a necessity for health.)

The sun is important in the garden as the source of all our energy. It causes the thermal currents in the air which lead to wind and rain. It drives the process of photosynthesis by which plants get hold of energy. It lights the day, and by lunar reflection the night. It gives us the warmth to survive.

Because the Earth moves relative to the sun each day, it appears that the sun crosses the sky. It's easy to imagine the sun rises in the east and sets in the west, but it does that only on the equinoxes (twice a year). In summer the sun rises and sets nearer the shade side pole (north in the northern hemisphere, south in the southern hemisphere) than in winter. The extremes are reached on 21 June and 21 December (give or take a day) each year. This means that every site has a varying sun sector throughout the year. Some space gets full sun every day, some is in full shade, and the area between gets a changing degree depending on the time of the season.

Plants should be placed accordingly. When designing tree plantings or structures such as sheds, the effect of the obstacle on shade can be measured by holding up a pole of the right height in sunshine and observing what the shadow does. Remember that the zenith of the sun (its highest point in the sky) varies with the season also, so longer shadows are cast by low winter sun and shorter ones in summer.

Solar gain is a measure of the ability of a particular material to absorb solar energy. There are some tricks for increasing solar gain. As light-coloured foliage, such as grey/silver leaved and some variegated plants reflects more light than darker leaves, this can be used to increase sunlight for other plants in a dark corner. These are also good friends in the battle for drought resistance, since the reflection of sunlight reduces water losses through evaporation (which is probably why such plants originally evolved).

Greenhouses are the obvious means of increasing the catch of solar power in the garden. Glass refracts (bends) sunlight, so that internal reflections are not easily lost back through the glass; thus a build up of heat takes place. This is particularly useful if greenhouses are positioned against buildings (the house, the chicken shed etc.) as the solar gain also reduces heat loss from the building (remember multiple uses here!).

Mulches can also be used to modify available sunlight. A winter mulch of straw, which is highly reflective, will increase light available to plants in the darker months of the year. A spring covering of black plastic will absorb heat and help the ground warm up more quickly.

Darker leaved plants radiate long wavelength (infrared) light at night from solar radiation stored in the daytime. So coniferous trees, in particular, have relatively warm perimeters on winter evenings. Ponds can also be used to reflect the low winter sun onto adjacent shade-side structures or plants.

Remember that in all climatic considerations the macro- and the micro-scale apply. What the weather does on the broad scale may vary enormously from what happens in small corners of the garden. This is especially true in town gardens where rows of buildings can cause wind tunnel effects, with sheltered and windy spots quite out of character with 'the prevailing wind direction'. Observe and act accordingly is the best advice.

Wind

Wind is probably thought of more as an enemy than a friend in the garden. Storms bring down trees, and bitter winter blasts kill off the cherished tender plants. Yet without wind plants would be weaker, their woody stems being (amongst other things) adaptations to the pressure of airflows. Without wind the air would stagnate, and disease would be rife. And without wind we should have no wind-borne pollination, and no windmills!

On most sites there is a prevailing wind direction, that is, a quarter of the compass from which the wind blows more constantly. There are usually other important wind directions to note, although at some times of year you can find the wind blowing from any direction.

In Britain, with its maritime climate, the prevailing wind is south-westerly,

Table 8 Drought-Resistant Plants

Some plants which resist drought.

Yarrow	*Achillea millefolium*	P
Onion family	*Allium* spp	A/P
Artemisia	*Artemisia* spp	P
Orachs	*Atriplex* spp	A
Broom	*Cytisus* spp etc.	P
Eucalyptus	*Eucalyptus gunnii*	P
Ice plant	*Mesembryanthemum crystallinum*	P
Lavender	*Lavendula* spp	P
Mulberry	*Morus* spp	P
Marjoram	*Origanum marjorana*	P
Pear	*Pyrus* spp	P
Holm oak	*Quercus ilex*	P
Rosemary	*Rosmarinus spp*	P
Santonila	*Santonila chamaecyparis*	P
Stonecrop	*Sedum* spp	P
House leek	*Sempervivens* spp	P
Tamarisk	*Tamarix* spp	P
New Zealand spinach	*Tetragonia tetragonioides*	P
Gorse	*Ulex europaeus*	P

following (or perhaps bringing) the warm ocean currents from the Caribbean. Yet the most damaging winds for the garden are the north-easterlies, which blow in chilling frosts, just at fruit blossom time. I recently saw a commercial nursery where after twenty-five years operation they'd eventually worked out that they needed a northerly windbreak just as much as their well-established south-westerly one. Although most aspects of sustainability are common sense, they're not necessarily obvious!

Wind direction will vary in other climates, so that continental North Americans, for instance, will be used to the cold northerlies blowing constantly down in winter across Canada, as the high pressure over the snowy Arctic tries to equalize with the lower pressure further south.

Accurate records can be obtained from local weather stations, fire services, local TV radio and newspapers, schools or well-trained neighbours. There are other tell-tale signs: trees are the best. They will tend to indicate the strength of the prevailing wind by the degree of list to the leeward. The storm-blown pine of the coast, lying nearly horizontally, is contrasted at one extreme with the straight stately parkland trees of mild protected areas on the other. If setting up garden in an unknown area this can be a useful indicator.

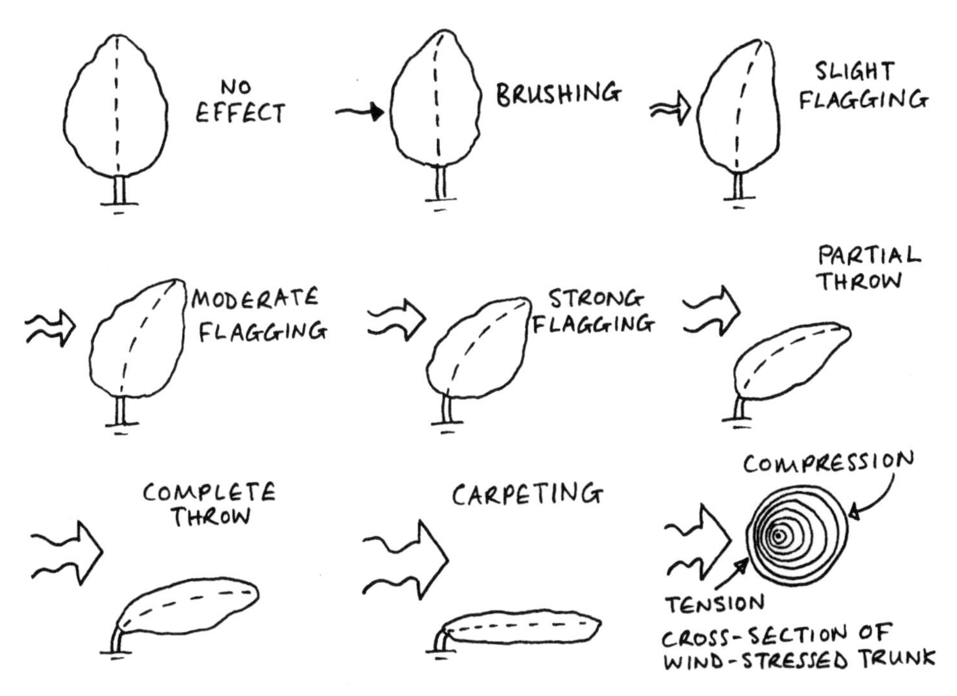

Trees as wind indicators.

Table 9 Wind Hardy Plants

Wind hardy plants can be used in their own right, or to 'nurse' more tender plants in their shelter.

Grey alder	*Alnus incana*	P
Sycamore	*Acer pseudoplanatus*	P
Monterey cypress	*Cupressus macrocarpa*	P
Escallonia	*Escallonia* spp	P
Beech	*Fagus sylvatica*	P
Poplar	*Populus* spp	P
Blackthorn	*Prunus spinosa*	P
Willows	*Salix* spp	P
Whitebeam	*Sorbus aria*	P
Swedish whitebeam	*Sorbus intermedia*	P

Tender plants should be protected (at least during establishment) through the winter. Wind chill causes the effective temperature to drop, and so although the air temperature may be above freezing, plants get wind burnt in strong blows. Old brushwood, hurdles or fences can be used temporarily. Long-living yew trees used to be cut back to provide winter protection for tender fruit trees in walled gardens in Britain. A modern version for small plants would be to take plastic bottles, cut out the bottoms and use them as miniature cloches.

The best windbreaks actually allow the air through, thus slowing it down. A solid windbreak creates much turbulence in its lee and is counter-productive. A 'do-nothing' aid for the winter garden is to leave tall dead growth such as beans, corn and sunflower stubble or lupins standing through the cold season, as the wind speed at ground level is slowed down. This is very helpful for the germination of tender seedlings in the early spring. Remember, seedlings are affected by the temperature immediately above the ground, which may be quite different from what you feel at face height.

There are a number of useful garden devices which can be wind powered. Wind chimes are an attractive addition to the garden which also make good bird-scarers. There are several variations on this theme. Old cycle inner tubes stretched between two posts vibrate and frighten birds. The 'flapping hawk' can be simulated by painting the silhouette on a rotating triangular drum. Wind-powered water pumps or electricity generators (small scale) are realistic options in even the smallest garden.

Water

Water is the transport system for nutrients in the garden. Plant cells need their water content to convey food upwards from the roots, and down from the leaves, and to prevent wilting.

The presence of water is obvious when we see it flowing in a pond or stream. However, the best place to store water in the garden is in the soil or in the plants. Soils which have good clay content or plenty of humus are the best at holding onto water. The more biomass (living organisms) on the land,

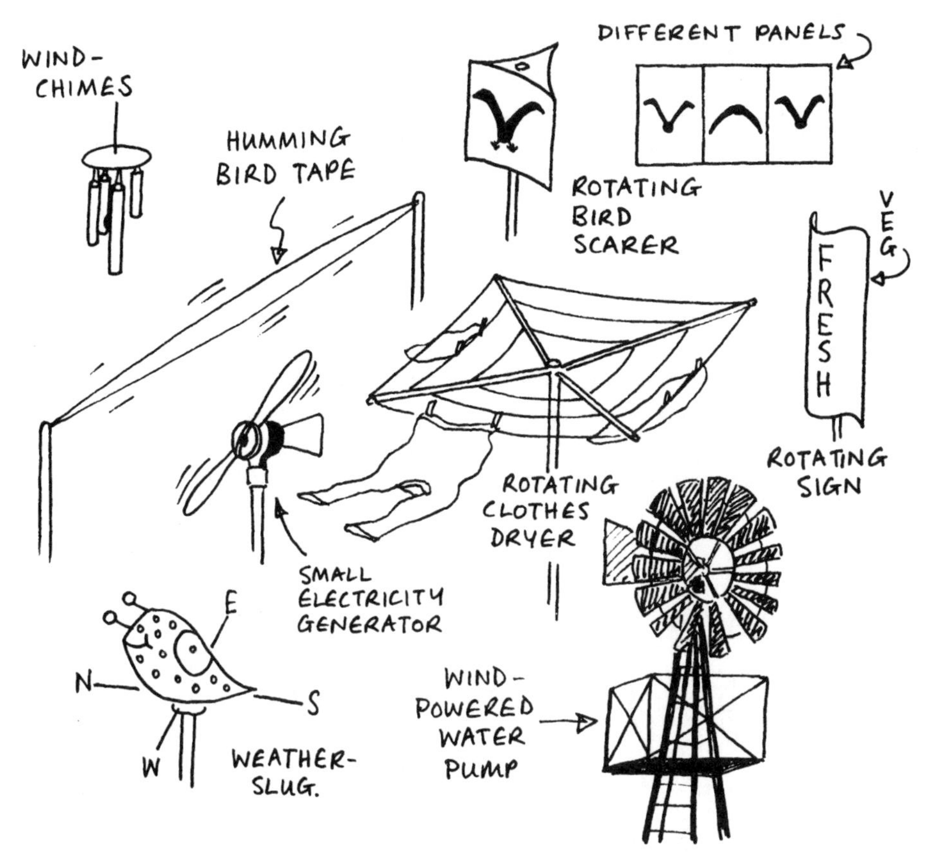

The sustainable gardener collects ideas for useful devices–remember, the wind is free!

the greater is the amount of water retained in the ecosystem. Some plants, such as succulents, have a particularly high water content, as they evolved to withstand drought conditions or desiccating environments like the salt spray of the sea's edge.

Water sources on each site vary, and include rainfall (or precipitation by other means), throughflow from any springs, streams or rivers, waste water from the house and the collection areas of any roofs, and underground supplies from aquifers or the water table, supplied by wells or pumps. In reality very few people have the luxury of natural water courses on their plot and will be looking to piped mains water for their supply. The majority of people live in urban or suburban situations, so we shall be paying particular attention to possibilities for the town garden.

The first statistic to note is the annual rainfall for your area; then to relate this to slope and shade in the garden. Shadier ground will be wetter, and steeper ground drier. But these are generalizations. Below dense hillside tree-cover the ground may be damper than in lightly shaded low-lying ground. The nature of the ground management will also affect how much of the precipitation infiltrates to plant root level. We shall look at ways of managing this in Chapter 8, 'Water in the Garden'.

As climate has become more erratic in recent years (which seems to be a global phenomenon) we should look to planning water resources carefully. It may be the most crucial limiting factor in the sustainability of our gardens.

Multiplying annual rainfall by floor area of the house will give the volume of water which can be captured from the roof of the building in a year. This will direct us to the volume of storage we need to provide to gain free water supplies for the garden. This also alleviates the stress on municipal water supply systems.

Just as much water comes out of a house as you put into it. At present the majority of people watch their waste

Drain pipe interceptor courtesy Blackwall Products.

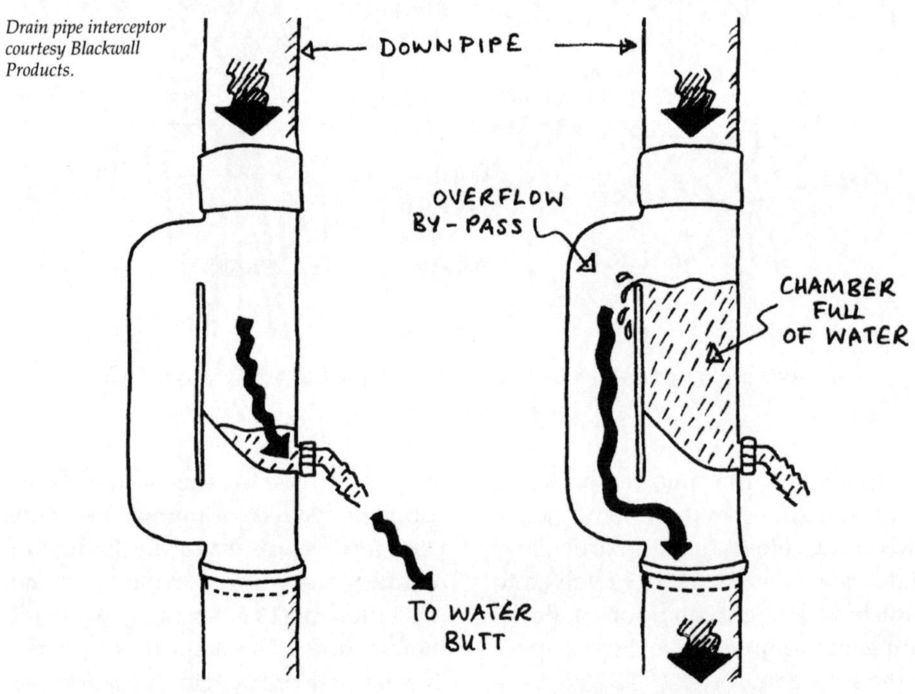

Catch the rain! Increasingly available, these interceptors make easy work of harvesting roof water.

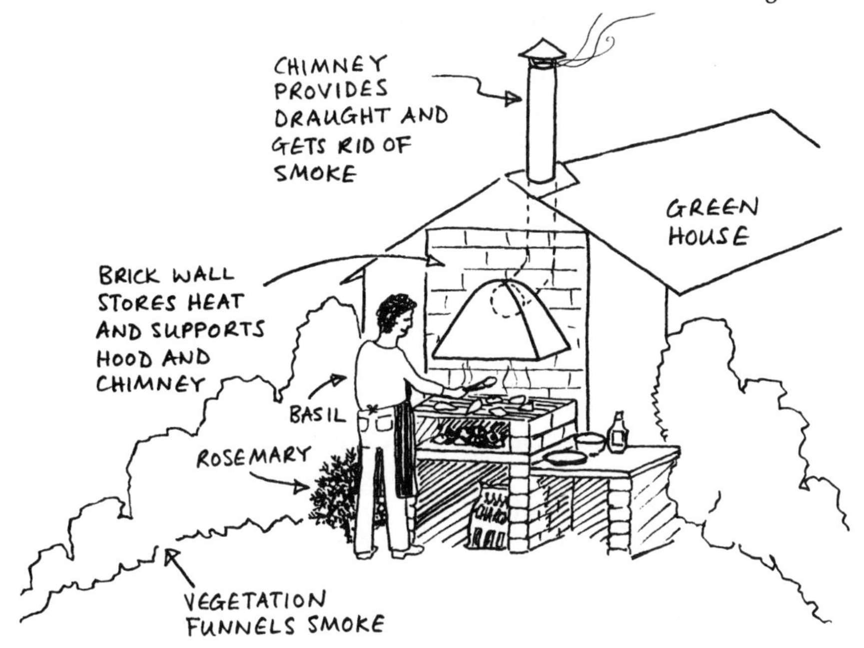

water disappear into disposal systems where the water is returned to the supply system, or to natural drains after partial treatment. Very little human sewage is left completely 'clean' after treatment.

We can devise systems for cleansing or reusing water on site which can add to the sustainability of our community by reducing the energy demand for this aspect of waste disposal. For instance, there is little point in flushing toilets with pristine spring water when slightly soapy water from the wash-hand basin would do just as well.

Up to 30 per cent of our urban environments is impenetrable to rainfall and so we can also look to paths and roadways as important collection points. Obviously we must use such water only in the certain knowledge that its quality is adequate for the uses to which we put it.

Fire ~

In temperate climates we tend to forget fire risk, as if arid areas alone suffer this threat. Fire is consuming and can devastate vast areas in hours. The tendency is for fire to run uphill, especially with the dry summer winds, so if fire risk is a consideration in your area, fire breaks should be left on the downhill and windward side of the house. Putting ponds in this direction can add to the safety factor. Note that some trees (like resinous pines with dry bracken undergrowth) constitute a high fire risk whereas others, such as mature deciduous trees, may be less so. House leeks (*Sempervivens* spp) are so called from the habit of growing them on the roofs of wooden houses to reduce the risk of fire spreading, as well as for their insulating properties.

Fire is also a useful commodity,

however, as it enables us to warm ourselves and to cook our food. External walls of buildings which back onto internal fireplaces and chimneys offer warm spots for bringing on more tender plants, such as grape vines. In addition, it is delightful to eat outdoors, so the provision of a barbecue spot with seating and tables for warmer weather is a nice way to bring the goddess of the hearth out into the garden. Appropriate placement involves consideration of wind and shade as well as trying to design smoke removal so that it annoys neither you nor your neighbours.

Energy Conservation ~

A well-designed garden is an exercise in total energy conservation. We want to do as little work as possible, to make it all productive, and waste as little as possible to minimize the pollution we create.

Labouring

- Are the tools stored near where they're used, and easily returned to storage points so you know where to find them?
- Are garbage/recycling (outward) and fuel/food (inward) stores handy for the kitchen door?
- Are green manures and compost/leaf heaps next to where they'll be used to minimize barrowing?
- Have you positioned labour-intensive plants for easy access? Place nearest the house the things that require attention most often, e.g. herbs for picking should be by the back door, and soft fruit close in, whereas potatoes can be further out.

Predators

- Do you have low-work strategies for dealing with predators?
- Are the points in the system where energy can be lost (fences/dams/greenhouse door etc.) regularly checked to prevent losses?

The Elements

- How well are things placed in the garden to take advantage of the sun, shade and wind, or to be protected from them if necessary?
- How well is the garden designed to offer seasonal successions to maximize productivity all year round?
- Are structures like food stores, greenhouses, and cold frames adequately designed to give maximum quality of output by keeping steady temperatures, or being easily vented if necessary?

Water

- Does watering take up a lot of time or has the garden been designed to be self-maintaining as much as possible? Include here systems for chickens and any other livestock.

It is also worth considering what inputs are provided from outside the garden, and to devise ways of producing them internally. For instance, can imported manure be replaced with green manures grown on site? If the garden has any waste which you currently export, consider whether there are ways in which that could be turned around for internal use.

Many people throw away woody cuttings from the garden, which would

Table 10 Insect Repellant Plants

Some plants deter unwanted insects and other pests.

Wormwood	*Artemisia absinthum*	P
Mugwort	*Artemisia vulgaris*	P
Lavender	*Lavendula* spp	P
Ox-eye daisy	*Leucanthemum vulgare*	P vs houseflies
Tobacco	*Nicotiana tabacum*	A
Fleabane	*Pulicoria dysenterica*	P burn for best effect
Rhubarb	*Rheum rhabarbarum*	P use leaves
Rosemary	*Rosmarinus officinalis*	P
Rue	*Ruta graveolens*	P
Mexican marigold	*Tagetes minuta*	HH A vs eelworm/couchgrass/ground elder
French marigold	*Tagetes patula*	TA vs whitefly/nematodes
Tansy	*Tanacetum vulgare*	P
Nasturtium	*Tropaeolum minus and major*	A attracts aphis (away from other plants)

make excellent compost or mulch. If you don't like the idea that they take longer to rot, a chipper is a good way of quickly reducing the hard material, and speeding up the whole process. The cost can always be shared among a group of neighbouring gardeners.

When laying out the garden, try to plan activities so that those which are done most frequently are nearest the centre of operations (that is, you and your home). Many folk lucky enough to have large gardens lose the vegetable plot out of sight at the bottom of the garden, having unproductive lawn and flowers by the back door. Interplanting of food plants and ornamentals can be highly attractive (indeed if treated right many vegetables are highly ornamental themselves) and proximity to the back door will ensure they get more frequent attention and so give higher and healthier yields.

Slope and Contour

If everything is an asset, then the slope of your garden should be seen from that point of view.

Remember that if warm air rises, then cold air falls, and so frost has a tendency to run downhill. Make sure that if your garden is frost-prone you design your use of slope to take advantage of the self-draining properties of frost to keep it off your tender plants as much as possible.

Slope also offers natural irrigation opportunities. Since water tends to run downhill, it should be stored as near the top of the slope as possible, and allowed to flow slowly through the slope, rather than simply draining away.

Again, if your soils are heavy and retain too much water, then slope can be used to alleviate the problem. Banks

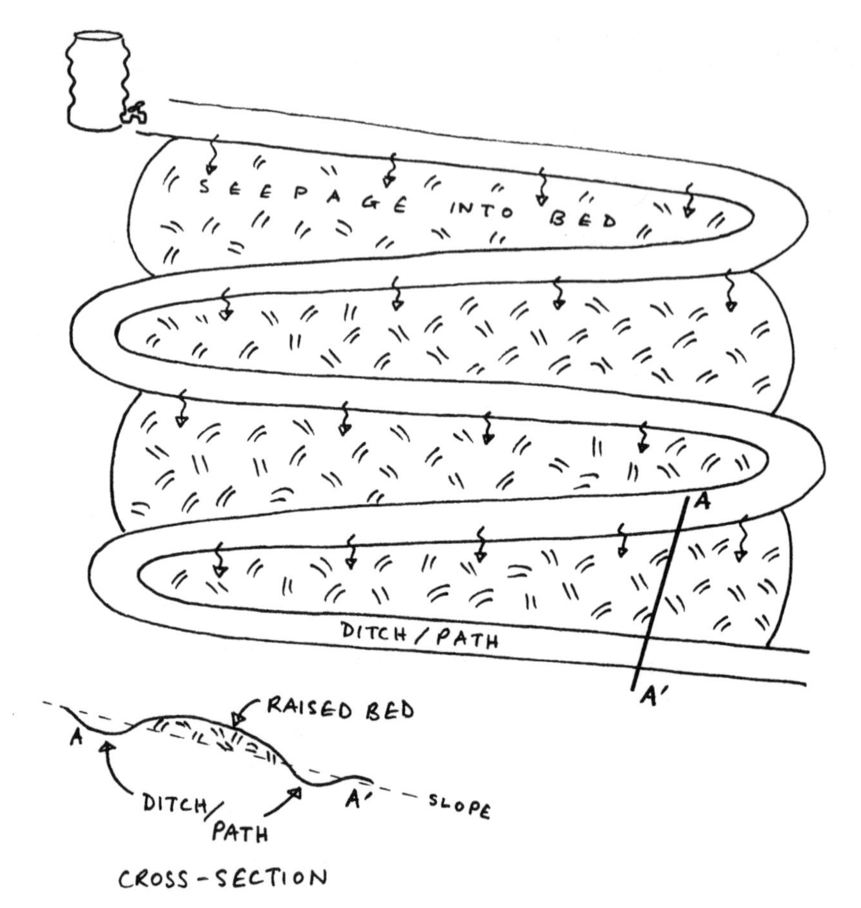

Here a water tank supplies a flow of water which runs over paths created slightly off contour. As its runs it seeps through to root level of the raised beds, watering a large section of garden effortlessly.

with a sunward aspect offer greater solar gain as they are angled to catch maximum warmth and light, but will also need protecting against long dry hot periods. Shade side slopes will be best for plants that like those conditions, and for cool spots in summer.

Whilst gardening on a slope sometimes seems like hard work, it is advantageous in that more light reaches individual plants. Certain crops, like grapes, thrive on free-draining sunward hillsides.

More easily workable areas can be made on slopes by the creation of terraces. There are various techniques for making terraces, covered in Chapter 6, 'Earth Shaping'. Terraces can be made 'on the contour' (i.e. level) or 'off the contour' (i.e. sloping gently downhill). 'On the contour' terraces are good for stopping downhill run-off of soil and water. 'Off the contour' terraces create water and frost drains and allow paths and water to move through maximum distances downslope. This has the advantage of increasing opportunities for harvesting the edge effect. The winding hillside path can be curved to create a varied micro-climate and many different growing opportunities.

Access

Access is something to design early in the layout of a garden, along with water systems. You don't want to have to fell carefully nurtured trees because they're planted where you subsequently decided to put the pathway. Also, much-travelled ground becomes compacted and therefore less suitable as growing space. Access lines can double as irrigation channels.

An effect of paths and roads is that, as they are frequently travelled, they extend the central zone of attention. So path edges become places to position frequent attention crops, such as salads, small fruit and herbs. Food collection time becomes an idle wander round the garden with a basket and the pleasure of checking the changing textures of the season.

Access should be designed to be wide enough, shallow enough and smooth enough for all purposes. That includes wheelchairs, wheelbarrows, pushchairs and prams, not just walkers and mountain goats. Surfaces should be as maintenance-free as possible. A good technique is to dig out topsoil and add it to adjacent beds. Lay down scrap polythene to prevent weeds growing through, contoured to provide a drainage camber at the path edge. Then layer the path with crushed rubble, chippings or bark, all of which drain well, are easily weeded and give a hard wearing surface. Fine chicken netting pegged over wood chips makes a path with good traction for wheelchairs. This technique also helps combat slipperiness on wooden steps and bridges when wet.

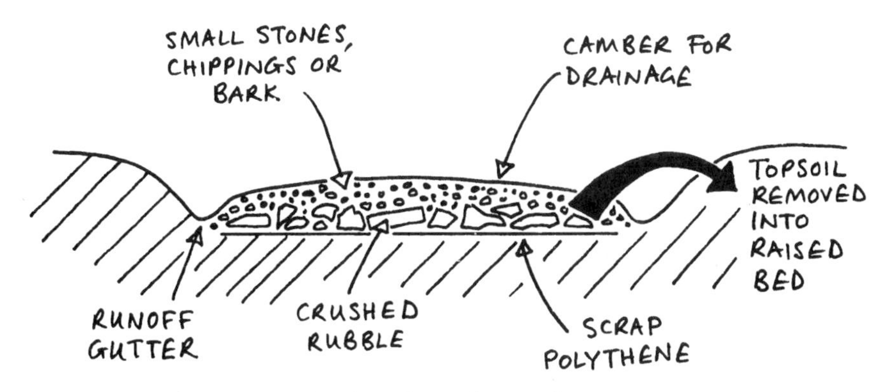

Section through a path.

five

Helpful Techniques

Nature in her bounty and abundance has provided for Man's sustenance and support with amazing prodigality.

J. SHOLTO DOUGLAS, *ALTERNATIVE FOODS*, 1978

The permaculture garden must be highly productive, but require as little work as possible. These needs may seem contradictory, but there are many techniques that can help. By practising the ideas outlined here, gardeners will increase their ability to develop such approaches for themselves. It will also become easier to spot similar ideas in action in other gardens, and to adapt them for your own use.

Minimizing Effort

The idea that gardening should be as little work as possible isn't just the lazy person's solution. It is an essential hallmark of understanding that ecological damage is usually the result of human intervention. Doing as little as possible is the best way to help the environment. This is nowhere more true than in the garden.

The ultimate example is to do nothing at all on your plot. What will happen? Assuming you start with a cultivated space, then initially the plants already there will carry on growing. Gradually weeds will take over. Why? Because it's their job. Weeds are plants especially adapted to certain conditions. They do not grow in isolation, but as communities. So the weeds that favour the edge of marsh land are very different to those you will find on a chalk upland or a clay vale.

If you know enough about plant likes and dislikes you can 'read' the soil conditions simply by observing what weeds are flourishing. The weeds that appear in the untended garden will be those adapted to the local soil and climatic conditions. Their particular place in the ecology of your garden is to cover bare soil (which is why they are so prolific), and feed the soil ready for the next stage in succession.

Nature is never static. In the wild every form of ground cover is in the process of changing and evolving. So the weedy meadow created by leaving a lawn uncut will become scrub in a matter of three or four years. Within ten or fifteen years the scrub will become woodland. In a human lifetime the woodland will become high forest.

A garden left to go to weeds will be highly productive–for wildlife. It may even yield wild fruit, nuts and herbs for your pot, but not in anything like the intensity that we generally expect to produce in our back yards. So most people want to manage their gardens. The 'do nothing' philosophy is, however, useful: it gives us time and space to observe how nature works if left to its own devices.

If you have the courage to do it, you can learn a lot about a new site by doing nothing for one year and simply observing how the seasons affect the plot differently. Learning where the frost lies late, or which corner catches late afternoon sun in summer and autumn may save you much grief in bad plantings later on.

Rushing in to 'tidy up' the garden may mean you remove something whose importance you don't fully understand. I recently moved a shrub, *Viburnum bodnantensee,* from a dark woodland edge where it couldn't be seen, to a sheltered spot by a wall. It withered badly in the winter. The woodland edge is much warmer than the side of a stone wall which radiates cold on frosty nights. Looking and thinking would have saved the plant.

A straggly old bush that's not particularly attractive may be an excellent windbreak. Digging over mature beds may damage the soil structure. Mowing the lawn too early may ruin a previous gardener's bulb planting, or deprive you of a wonderful crop of meadow flowers.

Lastly, when you do take action in the garden, observe how it feels. Is it hard work? Do you get back ache? Do you hate doing that particular job? Then there must be an easier way. And you are the right person to find it.

Perennial Crops

Plants fall into three main categories: annuals, biennials and perennials. Annuals grow from seed, reseed and die in a single year. Biennials take two (or sometimes three) years to complete the same cycle. Perennials live for a number of years; exactly how many depends on the species and site.

In designing the permaculture garden it is essential to give prominence to perennial plants. Choosing plants that live longer is obviously in line with any philosophy that says we need to plan for the long term.

Perennial crops are labour saving because you do not have to replant them every year. Also, because they

Nearly all herbs and many edible wild plants are also perennial.

occupy one place for a long time they have the additional benefit that they greatly aid soil fertility. By having the time to build deep root structures they mine the minerals way down in the ground which are essential to plant health.

They absorb these plant foods in solution through their roots and carry them up to the surface through the structure of their cells. This new-found food is made available to the topsoil in the form of leaffall. Some, of course, will be harvested as crops. Some more will be returned to the soil surface as mulch or compost from cuttings and kitchen scraps.

Permanent plants deter the inveterate digger, and so also build soil fertility by maintaining an open soil structure with plenty of pore space and humus. Both of these help the soil retain water and air. This is good for the soil health, and therefore for the health of plants. Selection of which perennials you plant is important: for example, a hungry and light-excluding hedge may well make it nigh on impossible to grow anything nearby!

Emphasizing perennials does not mean that the sustainable garden has no annual plants. It simply creates a more durable framework into which the annuals can be fitted.

Table 11 Some Edible Perennials

Edible perennials include all tree crops and bush fruit, as well as the following

Chives	*Allium* spp	P
Shallot	*Allium ascalonium*	P
Egyptian onions	*Allium cepa var. viviprium*	P
Welsh onion	*Allium fistulosum*	P
Bamboo	*Arundaria fastuosa*	P
Asparagus	*Asparagus officinalis*	P
Perennial broccoli	*Brassica oleracea botrytis asparagoides*	P
Globe artichoke	*Cynara scolymus*	P
New Jersey Tea	*Ceanothus americanus*	P
Jerusalem artichoke	*Helianthus tuberosus*	P
Lovage	*Levisticum officinale*	P
Common mallow	*Malva sylvestris*	P
Bamboo	*Phyllostachys* spp	P esp.P.sulphurea, P. verdi glaucescens
Japanese artichoke	*Stachys sieboldii*	P
New Zealand spinach	*Tetragonia tetragonioides*	P

Mulching

The best long-term strategy for soil fertility is to mulch the soil surface. This is the process of covering the bare soil. Mulching can be done with durable materials like black plastic, which is good at retaining the moisture in the soil and absorbing heat. This helps the ground warm quickly in the spring to bring on young plants. However, it does nothing to feed the soil. Mulching with organic material which will rot down and feed the soil life, has more long-term impact.

It is counter-productive to dig-in the mulched material. Living soil has large populations of worms and other tiny creatures which thrive by carrying this detritus underground and digesting it in their own life cycles until it becomes humus. Think of mulching as a piece of

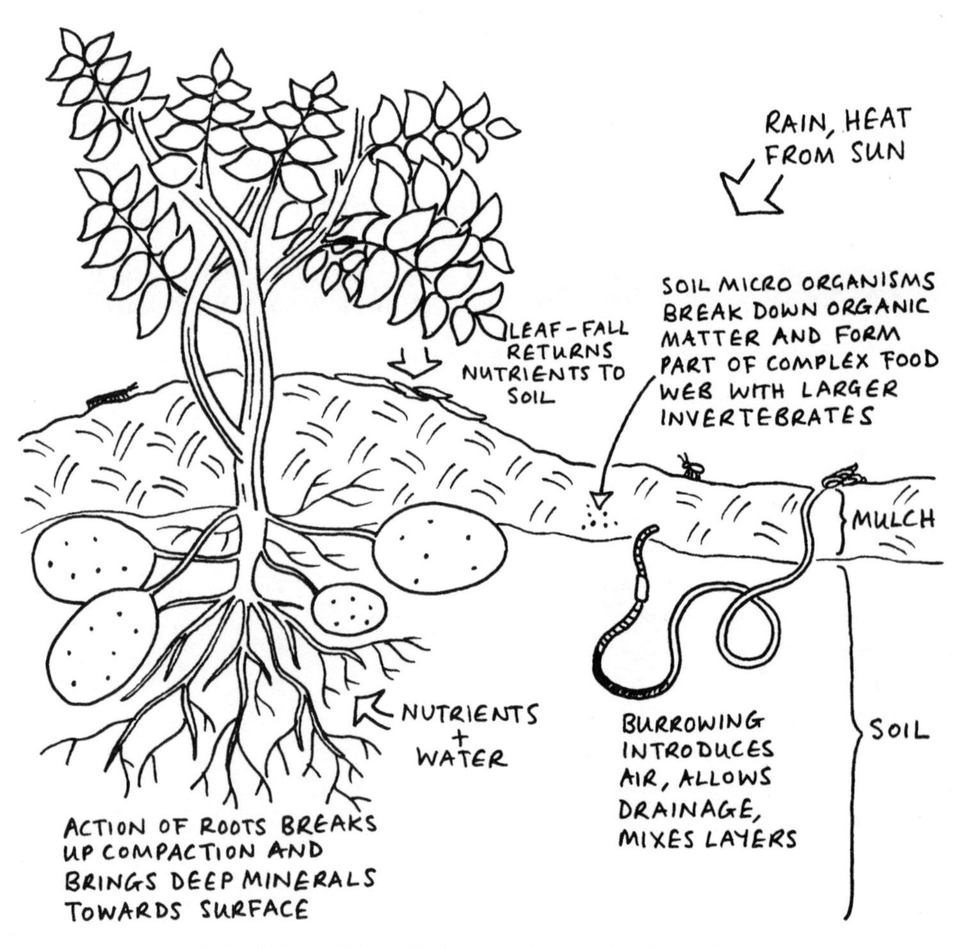

A healthy soil is a living cycle of interdependence

work which emulates what trees do in autumn, simply scattering their rich remains on the floor of the wood. It's as if we are feeding the soil life. You don't force-feed a dog if you want it to perform well, you let it come to the dish. The same applies to soil life.

Mulching protects the ground from evaporation in hot weather. So as well as increasing water retention through humus development, it has another advantage in the struggle to provide sufficient moisture for plants. This reduces the labour of watering the garden. It has the added benefit of reducing the need for precious treated water to be pumped onto your plot. Many cool climate countries have suffered severe droughts in recent years, so anything that reduces the need for artificial watering has got to be good!

Should the weather turn the other way and offer torrential rain, the mulch then protects the soil structure from damage and prevents erosion. Its fibrous nature is ideal for soaking up excess water.

At the same time, mulch tends to keep the soil temperature at a more even level than would be the case in bare earth. This is particularly important for giving consistent growing conditions for young plants.

Materials suitable for mulching include straw, grass clippings, leafmould, and chipped bark. In practice any organic matter can be used. Some things are more sightly than others. So if, for instance, kitchen scraps are being returned straight to the soil surface, they can always be covered up with a handful of cut grass to keep things looking tidy.

Some people find that mulching tends to attract slugs or other pests. The mulch itself will not do this. If the beasties are present they will make use of the cover, but it is the overall conditions in the garden which determine the presence of pests. Slugs tend to indicate moist conditions and acidity. Adequate water availability and slight acidity are excellent pointers for good growing conditions, however; most vegetables will thrive here. If the garden has a balanced population of plants and creatures then slugs will do little harm. (See also Chapter 3, p. **43**).

Plenty of herbs and flowers scattered throughout the plot tend to keep things healthy by attracting a range of wildlife (including insects) which keep the cycles of nature in balance. A vegetable plot which has no flowers will tend to be more subject to pests, just as will acres of unrelieved rose bushes. The bigger the target, the easier it is for disease or unwelcome visitors to strike!

However, mulching systems as described still involve lots of work collecting and applying the materials. Even if it is better for the soil (and less back-breaking) than digging and weeding, there's still a goal beyond that: to make the self-mulching garden.

Ground Cover

Look upon mulching as a stepping stone to the creation of continuous living ground cover. Aim to build a garden plan that has every piece of ground covered in growing plants.

Use applied mulching:

- to protect ground that has been freshly cleared
- to suppress weeds when making beds on previously uncultivated ground
- to build soil structure
- to protect and feed perennials which die back in winter
- to protect soil between annual crop rotations.

The principle of continuous ground cover is an important one, as we saw when we looked at mulching. It prevents unnecessary water loss by evaporation from the soil's surface, making the earth more resistant to the effects of drought. It also prevents run-off in heavy rainfall and erosion of dry soil in strong winds. Soil in nature doesn't leave itself bare, so why should the gardener?

A plot with living ground cover will produce a higher yield than one with 'dead' mulch. So, for example, strawberries are a quick self-spreading live mulch which yield fruit. A young strawberry bed might be planted up into a straw mulch. As the bed develops the straw rots into the soil, but the growing strawberry plants cover the surface instead.

Clover and other legumes fix nitrogen whilst keeping the soil protected. Any ground cover plant that gets out of hand can just be cut back with shears and left as green manure. On a plot which is to be left bare for a while, a green manure can be sown to feed and protect the soil meanwhile.

After a short time of working with soils in this way, weeded bare soils start to look like incipient deserts. The barrenness of the compulsively weeded garden has a machine-like geometry which becomes unattractive. The first day you bury your garden in four inches of straw or hay you may think it looks an unholy mess. With time, the benefits speak for themselves, and your ideas of tidiness change. Besides, the initial mess doesn't last beyond the first year or two, by which time the ground area is well planted up.

A friend told me he'd 'put his sward into soil build mode', by which he meant he'd stopped mowing the lawn. The natural meadow of wild flowers followed by mulched beds as he got the time to make them was far more attractive and fruitful than ever a billiard table lawn could be.

Love your soil and it will repay you well.

Get a garden! What kind you may get matters not,
Though the soil be light, friable, sandy and hot,
Or alternatively heavy and rich with stiff clay;
Let it lie on a hill, or slope gently away,
To the level, or sink in an overgrown dell–
Don't despair, it will serve to grow vegetables well!

WALAFRID-STRABO, *HORTULUS* (*DE CULTURA HORTORUM*), 9TH CENTURY

Rotations and Successions

Rotation is the principle of never keeping the same type of crop in the same place. It is done for soil health, and most rotations are practised on a three- or four-year cycle. Potatoes, beans, *Brassicas* and root crops might be planted in succeeding years. By dividing the plot in four and having each quarter for one of these in phased rotation, you can have a harvest of each crop every year. The rotated plots are less likely to become exhausted of specific nutrients, and less likely to attract disease or pests.

In organic growing it is well established lore that rotations are good for the ground. This simply means that we don't allow one type of plant to grow on the same part of the garden year after year. Mostly it refers to annual plants, although it is worth applying on the less frequent occasions when perennials come to the end of their lives. Don't, for instance, replace an apple tree with another apple tree.

So, don't follow potatoes with potatoes. When the cabbages are finished, don't plant another *brassica*. When replacing exhausted raspberry canes, find a different part of the garden for the new plants. There are several good reasons for this approach.

When the same plant keeps popping up in the same place it attracts pests and diseases. Some of these are extremely difficult to eradicate. Spores of club root (which affects *Brassicas*) live easily in the soil for eight years or more, so you don't want to get them in the first place.

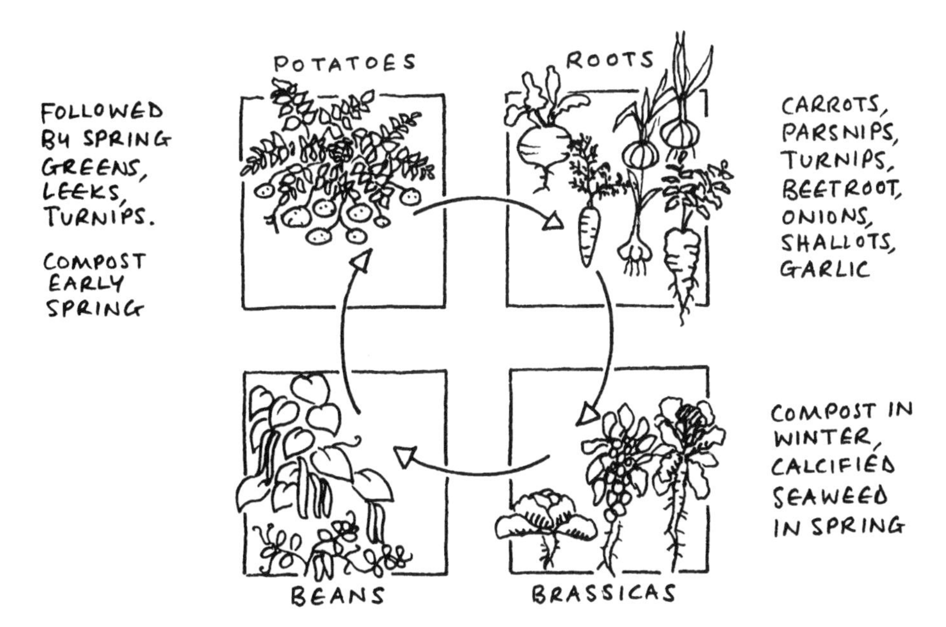

Traditional organic rotations.

Secondly, each plant requires a certain balance of nutrients. Repetition of that plant will exhaust the soil's reserves of those nutrients. Potatoes, for example, are very hungry for potash. Repeated harvests will reduce in volume and quality as available potash in the soil is exhausted. Certain plants put back goodness into the soil, so balanced planning can mean one crop contributes to the feeding of the following crop.

However, if, as we say, nature is our model, there's a question worth asking. Does nature operate rotations? No it doesn't. At least not in the way that organic gardeners practise them.

In nature each plant is adapted to a specific role. Each piece of ground is continually changing. The way in which the flora builds up is known as a succession. Take a bare ploughed field out of production and what happens? First of all weeds come in. Their seeds are already in the soil, and more arrive as wildlife or the wind bring them. The first weeds are prolific ground cover plants, such as clovers, chickweed and fumitories.

Next, the deep rooting soil reconditioners like thistle and dock establish. Soon these are followed by the first scrub–blackberries, elder, and thorn bushes might be typical.

In time the scrubland turns to woodland. Without human intervention most of Europe would be deep forest. Some areas favour savannah or prairie, whilst towards the poles and on high mountains tundra and alpine, ecologies thrive. Yet even a mature forest is not a static situation. The idea that a given environment will 'climax' as deciduous forest is another attempt to freeze-frame the process of nature. In time even oak trees die and give way to a grassy clearing, or revert to scrub.

Now you may be wondering how you are going to let your hundred square metre back yard succeed to high forest. Don't worry, it's not being suggested that you should give up your management of the garden! Whilst it's one possible solution if you want lots of wildlife and don't mind the neighbours complaining, we are suggesting a system of management. You design the kind of health offered by rotations through a managed system of succession.

The degree to which you can do this depends on what yield you want from your plot, and how much space you've got to achieve it.

This example would particularly suit a large rambling garden of the kind some Victorian town houses boast. Often these plots are rampant headaches to their owners who struggle to keep a wilderness under control.

Year 1

Convert old lawn (from three to ten metres square) by application of deep sheet mulch. Crops: spring-planted early potatoes followed by mustard, cut as green manure. In autumn plant with overwintering broad beans.

Year 2

Rake back any remaining mulch from around young bean plants and

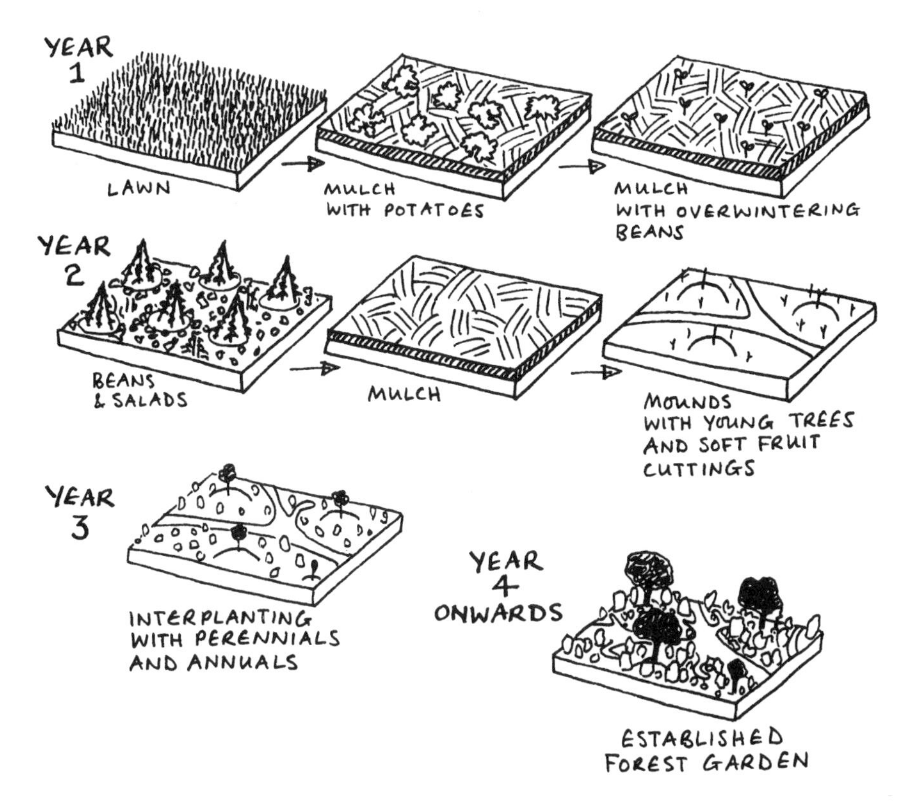

Managed succession develops a high-productivity, low-maintenance plot.

broadcast sow with hardy salad mixture–marigold, 'All year round' lettuce, endive, land cress, radish etc. Respread shallow mulch. Allow beans and salad seeds to grow up through this. Harvest beans in early summer (when ready), and salads on 'cut and come again' basis.

Plan fruit tree planting for autumn, and order chosen varieties. At end of summer, clear remaining crops, pull all weeds and save any plants which may be reused. Put in a nurse bed if necessary. Spread mulch or sow mustard until ready to plant trees. In autumn plant fruit trees at appropriate spacings. Buy in as 1+1 year whips. For small cultivars, spacing may be at three metres. Mound tree planting spots to provide raised beds, and lay out pathways between these. Use soil taken from paths (which are also drains) to build up beds. Interplant between trees with soft fruit bushes. These can be put in as cuttings. The ground is now quite open, as the trees are very young. This autumn, plant (or replant) strawberries, land cress, clover, daisies, and herbs as permanent ground cover. Plant out spaces with over-wintering greens. Mulch round all plants.

Year 3

Look to develop the beds towards a forest garden. That is, productive trees with interplantings of food crops and sufficient legumes and green manure plants that the garden becomes self-feeding. Perpetual spinach will attract beneficial insects, provide vitamin-rich greens, and any excess can be cut and left to lie as green manure. Brooms, gorse and lupins are all attractive perennials which can have showy flowers which also fix nitrogen and feed the surrounding plants. If they get out of hand, cut them back and use the clippings as green manure, by mulching back onto the plot.

Keep some spaces between the perennials for annuals, e.g. root crops, salads, or *Brassicas*, but try to grow these in small mixed plantings rather than large single crop beds. Make sure there are plenty of flowering plants and herbs to maintain garden (and human) health.

Year 4 Onwards

The trees may now be two to three metres tall (depending whether and how you prune them) and growing. As they spread out, keep growing green leafy vegetables and salads at the drip line (i.e. where rain falls from the outer edge of the branches). If your garden is large enough to move on to another area for other food crops, you may let the forest garden succeed into a semi-wilderness where you simply gather the wild produce of perennial plants and self-sowing annuals. Alternatively this could become a chicken fodder garden planned with plants for feeding free range hens. The only limit is your imagination.

For maintenance keep a supply of compost, small cuttings, raked leaves etc. to mulch any bare spaces, or suppress unwanted weeds. An annual pulling of persistent weeds in early spring or late summer should be sufficient to keep the garden developing.

Raw Food ~

If you garden to grow food, then no gardening book can ignore the overlap with the kitchen. One way in which your use of the garden can become more sustainable is to look at the potential for raw food.

There are two advantages to eating raw food. Firstly, as there's no cooking time, then it's generally quicker to prepare, and there's little energy expended in making the meal. Secondly, raw food can be an extremely healthy diet, in part, or whole.

Food loses nutritional value from the moment it is picked. So food picked fresh from the garden and eaten immediately has the highest possible value. Cooking breaks down important dietary material; exactly how much depends upon the food and the method of cooking. So, if you boil vegetables for twenty to thirty minutes and throw away the water, you are also throwing away a deal of goodness. Steaming for a shorter period of time keeps the texture and flavour of the produce. Eating raw food is even better.

Many of today's main crop vegetables

Table 12 Plants for Eating Raw

In addition to salads, edible flowers and most herbs

Allium family	*Allium* spp	A grated bulbs/chopped leaves
Cabbage family	*Brassica* spp	A Grated stalk/chopped leaves
Mustard & cress	*Brassica napus*	A Sprouting seeds
Chick peas	*Cicer arietinum*	A Sprouting seeds
Carrot	*Daucus carota*	A Grated root
Lentils	*Lens esculenta*	A Sprouting seeds
Sunflower seeds	*Helianthus* spp	A Sprouting seeds
Barley	*Hordeum vulgare*	A Sprouting seeds
Alfalfa	*Medicago sativa*	A Sprouting seeds
Radish	*Raphanus sativa*	A Seed pods/grated root
Parsnip	*Pastinaca sativa*	A Grated root
Aduki beans	*Phaseolus angularis*	A Sprouting seeds
Mung beans	*Phaseolus mungo*	A Sprouting seeds
Scorzonera	*Scorzonera hispanica*	A Grated root
Rye	*Secale cereale*	A Sprouted seeds
Soya beans	*Soya max*	A Sprouting seeds
Peanuts	*Trichis hypogae*	TP Sprouting seeds
Fenugreek	*Trigonella foenum-graecum*	A Sprouting seeds
Wheat	*Triticum vulgare*	A Sprouting seeds
Broad beans	*Vicia faba*	A Young tops

were only introduced in the last two hundred years on a mass scale. Traditionally people ate a lot more in the way of salads. Not necessarily our present lettuce, tomato and cucumber, but a whole range of wild plants, grated roots and chopped herbs.

Robert Hart (*The Forest Garden*, *Forest Gardening*), Joy Larkcom (*The Salad Garden*), and Leslie and Susannah Kenton (*Raw Energy*) all give good testimony to the enormous range of raw food which can be conjured from a small space, and the health benefits of eating it (see Booklist for details). In addition, salads are delightful to arrange, turning the simplest meal into a work of art.

Placing Things Well ~

An important element in creating a garden that is 'healthy by design' is the art of placing things well. This means, first of all, making sure you minimize human effort! Planting labour-intensive crops–and storing garden tools–in places that are easily accessible, for example, not only makes life easier for you but will also encourage you to work in the garden even if you only have a few minutes to spare. It is also important to place plants where they will be happiest; preferably in groups in which they can work together– thereby reducing your workload. (See also Chapter 2, and Chapter 7, 'Storage'.)

Composting

Composting is the process of returning waste organic material back to the soil by pre-rotting it. There are several schools of thought on the matter, which is why you will find conflicting advice in various gardening books. The systems described usually all work–but in different ways.

Some organic gardening organizations recommend composting at high temperatures. This requires a bin with ventilation at the base but solid sides to retain heat while providing adequate oxygen. Degradation of plant material generates heat, and by keeping this heat in, composting can be relatively quick. In the process seeds of annual weeds are burnt off. There are some difficulties with the logic of this method.

Most organic gardeners are opposed to bonfires in the garden, on the grounds that they are polluting and wasteful of useful compostable material. In practice is a hot compost heap not just a slow bonfire? Secondly, a compost heap will have its contents returned to the soil. Organic soil is a living myriad of beneficial creatures working away at feeding our plants. Clearly organisms that enjoy a warm compost heap are going to die as soon as they get turfed out onto the (relatively) cold ground. There will be a feed for plant life on their remains, but no increase in *soil* life from the addition.

Compost tumblers are a variation on this system which give quick results, a hygienic containment of kitchen scraps, and may be particularly convenient for small gardens. You can make one yourself, or buy them relatively cheaply.

The next approach might be described as the cool compost heap. There is a very good description of one version in Kenneth Dalziel O'Brien's book *Veganic Gardening*; see Booklist. Here the prime concern is to prove that healthy gardening is possible without animal inputs–so no imported manure. This fits in very well with the idea of a sustainable garden. Whilst I wouldn't deter anyone from using any available ethical products to get a garden established, I would encourage the would-be permaculture gardener to look at building a system which is self-fertile. Of course, vegans (who consume no animal products at all) would say that the use of animal products *was* unethical; whatever your views, the system proposed in veganic gardening is highly sustainable.

The 'cool' compost heap is made of layers of vegetable waste, green manure and earth. As it is cooler it takes longer to rot, and may not burn off all seeds. However, the compost is combined with a no-dig system. As Dalziel O'Brien points out, it is digging which releases buried weed seeds to the light and provokes germination. Weeds in a no-dig/mulch system are easily controlled within a couple of years of establishment.

In my experience the cool compost heap works well, and can take very rough material. The ideal is to build two heaps adjacent. When one is broken down to spread, any undigested material (woody stems etc.) is recycled into the new heap. The fine fibrous remains are mulched onto beds. If you are planning to sow seed after applying fresh compost it is best to leave a gap of two weeks; this gives time for any weed seeds in the mulch to germinate and be hoed out.

A moist, hot compost heap will attract brandling worms which help in its decomposition. Brandlings will not, however, live in the soil. A cool compost heap can be managed with earthworms, who will be added to the soil along with the finished compost, where they will carry on working for everyone's benefit.

A third and increasingly popular compost conversion option is to use worm bins. The compost is fed in layers into a bin which has been primed with worm eggs, or live worms. For most efficient conversion the bin should be air vented at the top and drain at the bottom. It should be kept warm in winter, either by being in a shed, or by being insulated (say with straw or newspaper jacketed around it). This will help the worms stay active.

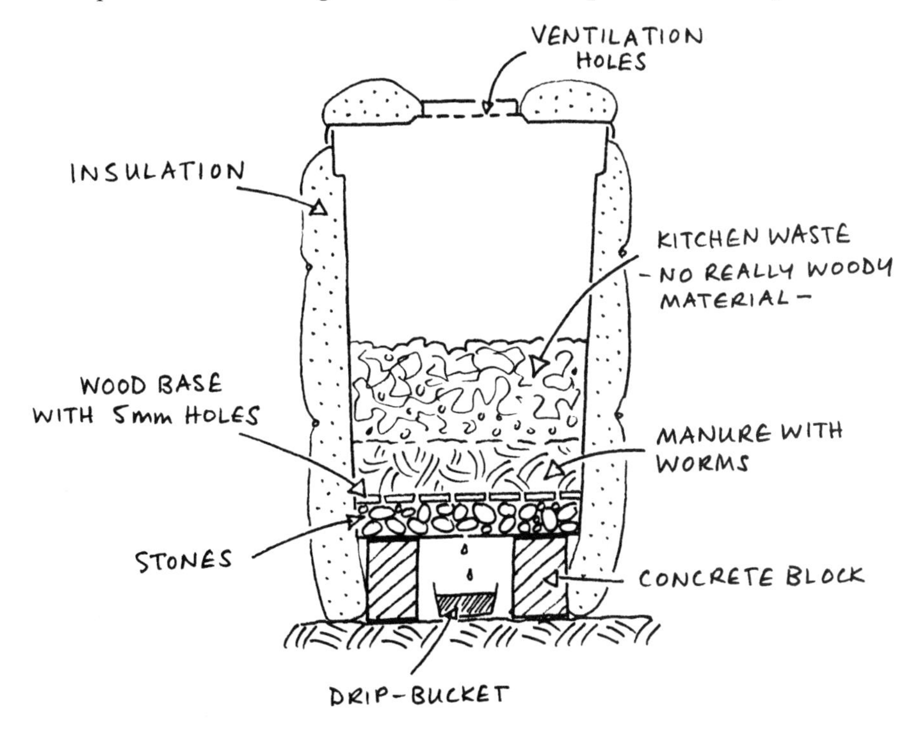

Worm bin.

The resultant compost, rich in worm casts, has a very high content of available minerals. It will contain worm eggs. If the worm farm is primed with earthworms (rather than brandlings, favoured by the commercial worm systems), then they will aid long-term soil fertility. Open University workers in Britain have found that the best worm populations were achieved by keeping them at airing-cupboard temperatures with a feed of shredded paper and brewer's yeast. I'm not sure I would go for that level of worm exploitation, myself. It does, however, indicate that moisture, air and warmth are helpful to keeping worms happy.

The final option worth mentioning here is that composting (a relatively labour-intensive activity) is completely unnecessary. The same end result will be achieved by returning all plant (or animal) wastes to the soil surface. Whilst it will take longer than the accelerated release of nutrients achieved by composting, all food content will in time be released back into the soil. The main drawbacks of this method are that it can be unsightly, and that it can attract unwanted pests.

This is a not unreasonable concern, as masses of fresh plant scraps spread around the garden are just the conditions to attract rats. Bearing in mind that a hungry rat is going to be even happier to snuggle down in the middle of a nice warm compost heap, the best advice about how to treat compost is that it's up to the individual to choose the system that suits them best.

Plants That Work For Us

So how can we make gardens that maintain their own ability to produce crops? The answer is to love weeds. Weeds are amongst the most useful plants. Not only are many of them edible, many more attractive, and all useful to creatures other than humans, but they tend to be very rich in minerals. That is because their roots draw up nutrients from the soil and below, which are released to improve the topsoil at leaffall.

Dynamic Accumulators

> *What is a weed? A plant whose virtues have not been discovered.*
>
> RALPH WALDO EMERSON (1803–82),
> *FORTUNE OF THE REPUBLIC*

'Dynamic accumulators' is a name given to plants with a particular ability to enrich the soil. To accumulate a particular mineral is characteristic of a plant which successfully lives on ground deficient in that nutrient. We have already remarked on how weeds are adapted to different conditions. Many of the processes by which these useful minerals are stored take place in the living soil below ground.

Plants that have chosen to colonize the ground betray its history. There are many examples: nettles indicate nitrogen-rich, recently occupied soils; willowherb comes in where fire has taken place; bracken may show much more ancient disturbance of the soil.

The plant's ability to mine and store

Table 13 Dynamic Accumulators

Some dynamic accumulators and what they store for us.

			Minerals
Chives	*Allium* spp	B	Na/Ca
Eyebright	*Anagallis arvensis*	P	S/K
Burdock	*Arctium minus*	A	Fe
Borage	*Borago officinalis*	A	Si/K
Caraway	*Carum carvi*	A	P
Chicory	*Chicorium intybus*	A	Ca/K
Carrot leaves	*Daucus carota*	A	Mg/K
Buckwheat	*Fagopyrum esculentum*	A	P
Cleavers	*Galium aparine*	A	Na/Ca
Alfalfa/lucerne	*Medicago sativa*	A	N/Fe
Bracken	*Pteridium aquilinum*	P	K/P/Mn/Fe/Cu/Co
Docks	*Rumex* spp	P	Ca/K/P/Fe
Chickweed	*Stellaria media*	A	K/P/Mn
Comfrey	*Symphytum officinale*	P	Si/N/Mg/Ca/K/Fe
Dandelion	*Taraxacum vulgare*	B	Na/Si/Mn/Ca/K/P/Fe/Cu
Clovers	*Trifolium* spp	P	N/P
Cattail	*Typha latifolia*	P	N

minerals and the conditions in which it chooses to flourish are directly connected. Bracken tends to appear on exhausted soils because it can accumulate its own supply of potash. When soils get very acid and sandy from loss of tree cover and overgrazing, then most of the potash is washed away ('leached out') in the rainfall. Bracken has deep roots which find and concentrate the missing nutrient. When it dies in the autumn it mulches the ground, replenishing the missing plant food. As a weed its job is to repair damaged soils.

Another example to look for is the dock family, potash accumulators adapted to damp, compacted soils. Comfrey is one of the most useful garden feeders because of its ability in this respect.

Plants have developed in this way to meet growing needs in certain conditions, but we can use those attributes to provide fertility in our gardens even when they have different prevailing soils.

This can be done either by growing the weeds which supply the plant food we want, or by harvesting the weeds elsewhere and scattering them as mulch. It is good to remember the principle of succession when doing this. The bracken, for instance will, in time, give way to leguminous plants, usually gorse or broom; these can live where there is little nitrogen in the soil. They continue the repair service to the next stage.

Legumes

Locally appropriate leguminous plants (e.g. gorse, broom, the pea family or acacias in a warmer climate) recolonize ground which is deficient in nitrogen, because they have the ability to fix nitrogen from air in the soil, in conjunction with specialized bacteria. These associations are known as *mycchorhizae.*

Legumes deserve consideration in their own right because nitrogen is the essential building block of living organisms. Humans need it to make protein in their bodies; they get it by eating either plant or animal protein. All plants in the garden need nitrogen to grow well (whether we eat them or not).

Modern agriculture has produced its phenomenal tonnage output on the basis of chemically produced soluble nitrates. Clearly it's not sustainable to continue this process. Oil (from which chemical nitrates are derived) is not unlimited. Chemical nitrate is highly soluble, which is why it feeds plants quickly. It also means, however, that a major proportion of the applied nitrate washes straight out of the soil, and pretty quickly, into our water supplies or back to the sea.

Traditional organic systems relied strongly on animal fertilizer to supply nitrogen, and to some degree on green manures (as dynamic accumulators) and intercrops of leguminous plants (usually clover, but also lupins and lucerne). We should not underestimate the nitrogen input from soil life in a living organic soil. The earthworms and various insects and bugs in the ground give considerable feed to the plant life when they die and return the minerals in their bodies to the soil. This system of land management had many advantages.

Today it is less certain that gardeners will have access to animal manures (unless human wastes are returned to the soil). Nor would it necessarily be desirable. Commercial pig manure must be avoided like the plague: pigs are fed growth promoters which contain copper and which remains in the soil. One fruit grower I know has found a great increase in tree pests and diseases since using this 'feed', which is actually, of course, a major pollutant.

Non-organic cow muck contains hormones and other pharmaceuticals, which will, however, break down in time. This kind of farmyard manure will be usable in the garden if rotted down for at least a year. Horse manure is an old gardeners' favourite. Again, watch out as this is often full of weed seeds, so seasonal rotting before spreading is a good idea.

How much better, though, to aim for true sustainability and try to provide nitrogen needs from the garden itself. Throw away that crutch! (I mean imported manure, of course.) Cultivate legumes!

Green Manures

One way to make fertility self managing is to allow weeds to do their job, which is soil repair. We may also consciously

Table 14 Legumes

Leguminous plants for the garden

Mimosa	*Acacia dealbata*	TP
Persian Acacia	*Albizia julibrissin*	TP
Kidney Vetch	*Anthyllis vulneraria*	P
Peanut	*Arachis hypogae*	A
Milk Vetches	*Astragalus* ssp	A
Pea Tree	*Caragana viscosa*	P
Judas Tree	*Cercis siliquastrum*	TP
Bladder Senna	*Colutea arborescens*	P
Crown Vetch	*Coronilla varia*	P
Crotolaria (US)	*Crotolaria* ssp	P
Brooms	*Cytisus* spp	P
Lablab (US)	*Dolichos lablab*	TP
Dorycnium	*Dorycnium hirsutum*	P
Goat's Rue	*Galega officinalis*	P
Brooms and greenweeds	*Genista* spp	P
Honey Locust (US)	*Gleditsia triacanthos*	P
Soy Bean	*Glycine max (syn Soya max)*	A
Licorice	*Glycyrrhiza lepidota*	P
Hedysarum	*Hedysarum multijugum*	P
Horseshoe Vetch	*Hippocrepis comosa*	P
Pigweed (US)	*Hoffmansegia densiflora*	P
Indigo	*Indigofera heterantha*	P
Adam's Laburnum	*Laburnocytisus adamii*	P
Laburnums	*Laburnum* spp	P
Peas and vetchlings	*Lathyrus* spp	A
Desmodium (US)	*Lespedeza bicolor*	P
Birdsfoot trefoils	*Lotus* spp	P
Lupines	*Lupinus* spp	A/P
Medicks and lucerne	*Medicago* spp	P
Melilots	*Melilotus* spp	P
Sainfoin	*Onobrychis viciifolia*	P
Restharrows	*Ononis* spp	P
Bird's Foots	*Ornithopus* spp	P
Milk Vetches	*Oxytropis* spp	P
Runner and dwarf beans	*Phaseolus* spp	A
Piptanthus	*Piptanthus laburnifolius*	P
Field Bean (US)	*Pisum arvensis*	A

Garden Pea	*Pisum sativum*	A
Black locust (US)	*Robinia pseudoacacia*	P
Colorado River Hemp (US)	*Sesbania macrocarpa*	P
Spanish Broom	*Spartium junceum*	P
Velvet Bean (US)	*Stizolobium deeringianum*	P
Asparagus Pea	*Tetragonolobus purpureus*	A
Thermopsis (US)	*Thermopsis gracilis*	P
Trefoils and clovers	*Trifolium* spp	A/P
Fenugreek	*Trigonella foenum-graecum*	A
Gorses	*Ulex* spp	P
Vetches and tares	*Vicia* spp	A/P
Broad Bean	*Vicia faba*	A
Wisterias	*Wisteria* spp	P

plant certain crops as green manures, that is, plants whose primary function is to feed the soil.

A typical low-work example of this is to interplant comfrey between currant bushes. The comfrey is self-mulching when it dies back in the autumn, but grows vigorously in the summer. Just pass by every six weeks or so and cut the comfrey down close to the ground. Lay the green material around the fruit bushes as you go. Instant manure and weed suppression combined.

There is an extensive list of green manures opposite, in Table 15. Remember that all plant wastes, be they kitchen scraps, shrubby cuttings, or lawn clippings are in effect green manure and should be used as such. Growing green manures for their own sake is advised whenever there would otherwise be a bare patch of ground in the garden.

Integration of Yield (Symbiotic Effects)

Moving away from mono-cropping ('this packet of seed is sufficient for one 30m row') to mixed plantings ('what would this go well with?') will increase yield in the garden.

The information to do this is something gardeners acquire continually, so we cannot hope to learn it all in one paragraph. The important thing here is to accept the principle, take some ideas for starters, and then develop your own 'portfolio' of what works well for you.

Symbiosis is the process in nature whereby living things work for mutual benefit. The *mycchorhizae* mentioned on p. **78** is an example. Two things get together, and they both do better as a result. The ideal marriage, if you like; by marrying together plants in the garden we should be able to ensure a better harvest for less work.

Comfrey with currants was an example above. Here are some more:

- Marrows with radishes and marigolds: good for colour and for beneficial insect attraction. Radishes crop early before the marrows take off, thus avoiding bare soil.
- Broad beans with apples, chervil and nasturtiums: nice colours again. Nasturtiums give good ground cover to conserve moisture which is needed by the beans and apples. They also attract blackfly away from the beans. Hoverfly is attracted by the chervil and then eats up the aphis. Lots of good green stuff to eat, and a year round growing 'guild.'

Table 15 Green Manures

A selection of green manure plants.

Borage	*Borago officinalis*	A
Buckwheat	*Fagopyrum esculentum*	A
Sunflower	*Helianthus annus*	A
Grazing rye	*Lolium perenne*	P
Lupin	*Lupinus* spp	A/P
Trefoil	*Medicago lupulina*	A
Phacelia	*Phacelia tanacetifolia*	A
Radish	*Raphanus sativa*	A
Mexican marigold	*Tagetes minuta*	A
Crimson clover	*Trifolium incarnatum*	A
Red clover	*Trifolium pratense*	P
Mustard	*Sinapsis alba*	A
Winter tares	*Vicia sativa*	A

The earth, that is sufficient.
WALT WHITMAN (1819–92),
STARTING FROM PANAMOUK

Productivity and convenience in the garden can both be increased with careful landscape design. In a natural garden the soil will be disturbed as little as possible after establishment, so these ideas are all for laying out initial designs. Of course it's fine to change a garden with time, but think of these techniques as ways to set up a carefully thought out garden which will endure and give years of pleasure.

Bed Types

Making beds in the garden is helpful from a management point of view. The growing areas and paths are separated so that topsoil doesn't get compacted by foot traffic, and paths can be built with surfaces which are weather and tread resistant.

Beds should be designed to take advantage of slope and aspect, to give maximum edge, and to minimize work. A flat rectangular garden offers much less opportunity for variations of light and shade, damp and dry areas, and complex associations of plants than one with many shaped sections. The use of raised and sunken areas and curved edges enhances the range of plants which can find a niche.

In contrast, a steeply sloping garden can be difficult to work and lead to topsoil and water loss through run-off. This condition can be greatly improved by terracing, which can also create beds which are easily worked without bending. This is particularly beneficial where a garden is designed for use by people with disabilities.

Learning different ways of laying out beds can, therefore, help increase productivity *and* reduce backache. We've already had a look at sheet mulching, the German Mound and raised beds based on central trees (Chapter 3) and at the principle of maximizing edge (Chapter 2). If we combine these various ideas, then many shapes and sizes of bed can be produced.

Raised Beds

I leave the whole plot to be baked, like a bun,
By the breath of the south-wind and the heat of the sun,
Only; lest the soil slip and drift out of its place,
With four pieces of timber I edge the whole place,
Then heap the bed up on a gentle incline.
Next, I rake til the surface is powdered and fine;
And lastly to make its fertility sure,
I impose a thick mulch of well rotted manure.
And now–a few vegetable seeds let us sow,
And watch how the older perennials grow!

WALAFRID-STRABO, *HORTULUS* (*DE CULTURA HORTORUM*), 9TH CENTURY

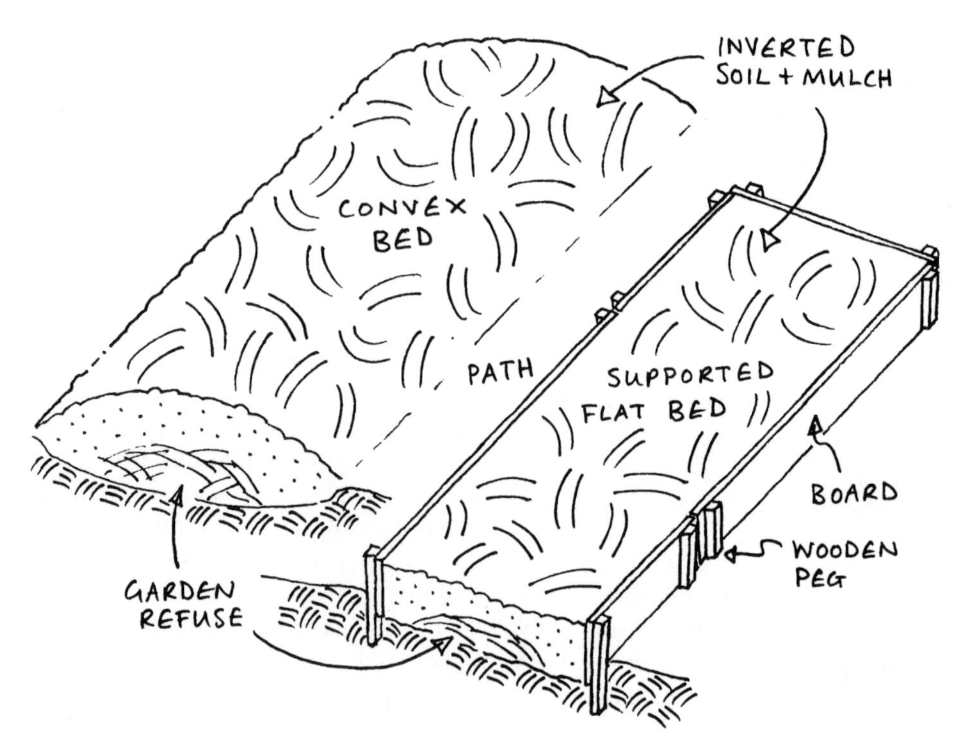

Raised beds offer greater growing area, more choice of micro-climate, and are more interesting to look at than growing on the flat.

A favourite amongst no-dig gardeners is the straight raised bed. This gives the advantage of increased growing area (one and a half times that of a 'flat' bed) which is achieved by raised beds generally, but means that we can produce crops in small blocks of varieties to be harvested seasonally. For this reason it is often used by commercial organic vegetable producers.

Mark the area to be made with a line. Make the bed by double digging. Take off the turf and topsoil from the first metre and barrow to the far end of the bed. Dig the second spade's depth of topsoil, cover this with inverted turf from the next metre section of the bed and then with the underlying topsoil. In cross section the bed should be heaped. Carry on until the end of the bed is reached and then cover over the last metre of subsoil with the turf and topsoil from the barrow, brought from the far end.

Some slight raising is achieved by digging; additional organic material placed around the inverted turf layer will raise it further. Digging out surrounding pathways can be particularly helpful. This creates a core to the bed which offers a long-term nitrogen release. When the bed is complete, maintain it by mowing the grass paths alongside, keeping the edges clipped, and by digging a gutter along each edge. Fine soil will accumulate in this small trench as a result of run off from the slope, and as birds feed amongst the plants. Continuous mulching between the plants will help this process, which develops a rich humic layer which is dug and spread back on the surface of the bed.

The gutter effect means that rain infiltrates to root level very quickly, so that evaporation is minimized and plants quickly get the benefit of any rain in dry periods. It also means that in heavy rainfall run-off of both water and topsoil is trapped and kept in the bed where it is needed. If the beds are run in an east-west direction, then they have a sunny and a shady side. Plants can either be planted to take advantage of their preference for differing levels of heat, or taller plants can be planted on the shade side so as not to shade out smaller plants which occupy the sun side.

Raised beds should be no wider than can be easily reached to the mid-point, so that it is never necessary to walk on them, giving maximum soil life and minimum compaction.

A simpler method of raised beds used by many organic gardeners is to edge the beds with old boards secured upright with stakes. Inevitably this gives rectangular beds. It shares the advantages of deepening topsoil and keeping beds free from compaction, whilst meeting a certain standard of 'neatness'. The soil will cool more quickly in the autumn/winter and warm more quickly in the summer because the bed is raised up into the air.

Another stage of development of this idea is to make the beds curved rather than straight, and to see what is the maximum bed-to-path ratio you can fit in a given area. You might prefer just to make interesting shapes rather than being mathematical about it. Most

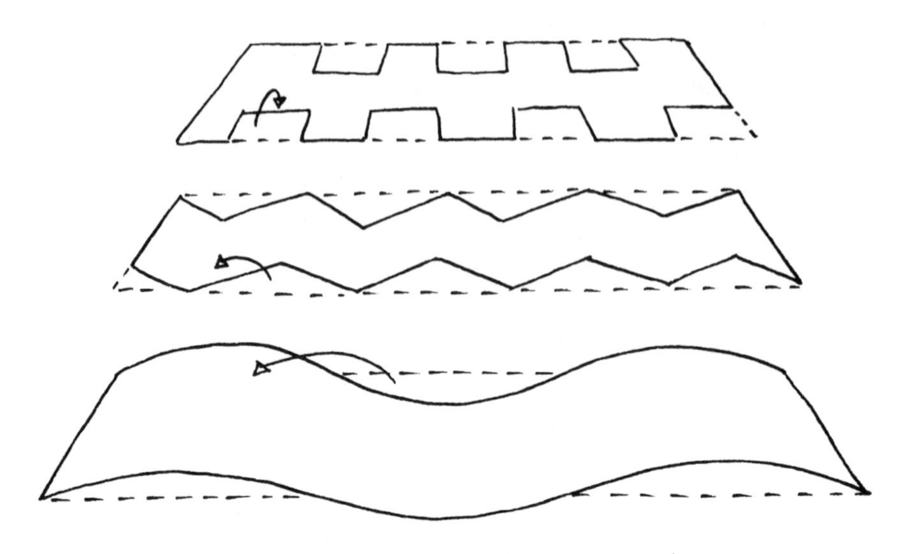

Edge and area are maximized with shaped raised beds.

people who try this method find it undesirable to return to square flat shapes in their garden. One reason is that it turns a rectangular flat space (which is most gardeners' starting point) into something attractive with new vistas at each turn, mysterious corners, and plays of light and shade which entice the viewer into the garden. A nice contrast to the intimidating feeling of 'Oh no, it's time to do the weeding'.

Ruth Stout Method

A woman of agricultural stock in the Eastern United States, Ruth Stout developed her method after the annual slog of piling horse manure onto the vegetable plot wore her out. She reasoned that if horses eat hay, you'd be better off using hay for manure as that had all the soil feed value of manure, but with the horse taken out! It's rather similar to the sheet mulch method except the sole mulch is a 10–20cm (4–8in) blanket of hay.

This is continually added to year on year, and the result is fantastic populations of soil life busily turning the mulch into topsoil. The desirable end result of this for the gardener is lots of fertile ground for the plants to thrive in.

Minimum Tillage ~

In the permaculture garden we are looking for systems that work with minimum intervention. There are two reasons for this. Firstly, it's less work. Secondly, we are interfering with nature as little as possible. To make nature work for our benefit, we design and place things so that the fertile processes of the soil increase the productive capacity of our garden, rather than looking to imported materials to make the growing areas fertile.

In all soil constructs, look to build long-term surfaces which require little or no digging. Mulching is one way of feeding the living soil and suppressing weeds whilst causing minimal damage. It is possible to remove deep rooting plants by inserting a fork and gently lifting the ground, without inverting the soil. A firm grasp on the plant, with gentle pressure, should release the deep roots from the soil.

If beds are made so that they receive little or no foot traffic and are not machine cultivated, then the need to till the ground to alleviate compaction is removed. Another job eliminated!

For all the reasons discussed above we should dig (or otherwise disturb or 'till') our soil as little as possible.

What is appropriate depends on the nature of our soil at the outset. Old ploughed land or soil which has been repeatedly dug in the past may have a layer of 'pan'. That is, relatively impermeable soil at the depth to which it was cultivated. In European conditions it usually indicates a concentration of iron. This should be broken through before any further action is taken. In farming this is generally done with a single furrow 'subsoiler'. In a garden it can be fractured at close intervals with an iron bar and sledgehammer if necessary, or a stout spade if you're up to it.

Not many gardens will be so affected. A soil profile will give the necessary information, and also show the general degree of compaction and leaching in the soil. (See Chapter 11 for further information)

Most people think soils should be weed free. Weeds compete with our crops for nutrients. It is also considered normal to dig, to break up compaction and prepare the soil as seed bed.

Let's think about what nature does for a moment. Nowhere does it invert the soil as we do with a spade, except in the glacial periods where over hundreds of thousands of years new soils are formed by the crushing action of glaciation on the existing landscape. Not a good time to garden! Instead, in the warmer inter-glacial periods which make life possible for people, tree and plant roots are continually stirring and breaking open the soil. Small amounts of earth are passed through the gut of earthworms and left at the surface in worm casts. Likewise moles and other burrowers leave fine friable piles of well-worked soil outside their runs. This gentle aeration actually leaves soil which is rich in available nutrients and of excellent crumb quality.

When we bury surface soil life (much of it invisible to the naked eye), or expose soil life from a spade's depth down, we generally kill off these creatures either through suffocation or oxygenation. Their dead bodies yield up nitrogen, and a flush of growth of our crop shows a short term increase in fertility. This is a good tactic for establishing garden beds. But if we continually till the soil then we start to compact it, and we harm the long-term fertility of our plot, making necessary all the busy work of fertilizer application.

Try always to find ways of avoiding the spade if possible.

Rock Gardens

Certain species are adapted to growing in rocky conditions. Ranging from mosses through succulents and ground cover plants to hardy shrubs and some trees, these are generally plants which create soil and which are resistant to drought. They tend to have roots which break up the surface layers of rock and create fine mineral particles, which with the organic remains of the plants themselves are soil in its infancy.

New gardens with a high rubble content, or those on rocky terrain, such as granitic soils, suggest rock gardens as a natural usage of an available resource. Rocks which split horizontally along the grain (usually sedimentary in origin) are easier to build into walls than eroded or volcanic rocks. Consequently dry stone walls are more a feature of limestone and sandstone country, and earth and rubble walls

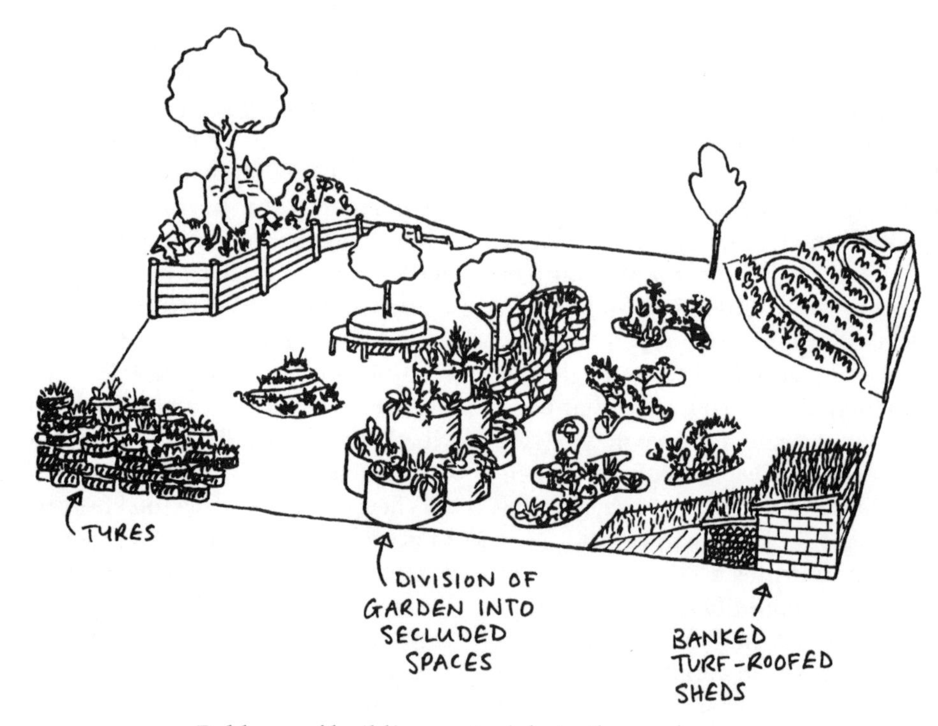

Bold use of building materials in the garden can greatly increase productivity. The dynamic visual effect can turn a flat plot into a maze of surprises.

(such as the Cornish 'hedge' which has bent many an unsuspecting tourist's car) are more suited to volcanic and river bed landscapes.

There are many attractive plants (some edible, such as stonecrop and iceplant) which are adapted to these conditions. They are mostly derived from seashore locations, where the plants have to survive dehydration from the salty environment, or areas of low rainfall, such as Southern Africa or the Mediterranean, or from alpine areas with thin soils. Because of this they can do well in the comparatively dry areas where there is more rock than soil.

An ideal use for rock gardens is to make them on the vertical walls supporting terraced areas. On a steep garden these vertical faces can occupy a significant area. By this method it becomes an attractive and productive growing area, rather than a blank stone wall. The flatter upper areas can support plants needing better soils, and the upright wall spaces can house the rockery plants.

Alpines, especially, may need careful understanding of their growing needs, so always get detailed advice before buying expensive (and possibly rare) plants for these areas. Many herbs (being derived from Mediterranean scrub terrains) will do well here if watered until established.

The herb spiral is a rock construct which increases edge, giving varied microclimate, and so makes high yields possible in a small area. Another example of being creative with space in a small garden!

Swales ~

Swales are shallow trenches cut into the hillside to encourage water to move along the contour, rather than to drain straight off. They are used by livestock farmers in dry conditions to keep as much water as possible *on* the hillside to help the grass grow. They can be just as useful in a sloping garden.

Swales often serve multiple purposes. They catch water running downhill and help it soak into the topsoil instead of running away. They trap water, leaf litter and loose soil which is running downhill. They can also act as road or pathways. In time they collect debris and fill up, forming terraces.

Swales are a good solution to improving water retention in soils on dry sloping sites, and decreasing topsoil erosion where that is a problem. Because of their tendency to trap organic matter and increase water reserves, they are an appropriate spot to plant crops which need wetter conditions, such as squashes.

Terracing ~

Anyone who has seen the terraces of South East Asia will have marvelled at how productive they have made the land. The need to use terracing to water rice crops makes them seem more appropriate in that part of the world, yet terracing also has an ancient history in many cool climate countries where it can be just as beneficial.

What do terraces do for us?

More gain, less pain.

- they increase the growing area over a sloping hillside
- they can take the back breaking out of working on the slope
- they slow down or stop erosion and water loss
- they make it easier to move about and work

Growing on a steep bank can be very difficult. You slip and slide everywhere, and end up with terrible backache. Paths and terraces bringing the beds to waist height up the slope can do away with all the sweat and tears.

Terraces can be produced in rough and ready fashion very quickly, or with engineered perfection, slowly and expensively. How you make a terrace will depend on your resources. Those beauties in Indonesia and China probably took thousands of years to develop, and they still need whole villages to maintain them.

We recommend you go for an easy method of establishment, and look to make something more substantial in time. Here's one method that may work for you.

Dig holes in lines slightly off the contour, into which posts are placed, and then hammered home. This way each bed will drain along the hillside. On the downhill side nail stout boards, starting at the top and working down. 10cm (4in) square is a good size for posts, and 3 x 15cm (1 1/4 x 6in) for the boards. Length will depend on the steepness of the slope and how far apart you put the posts. Obviously the whole

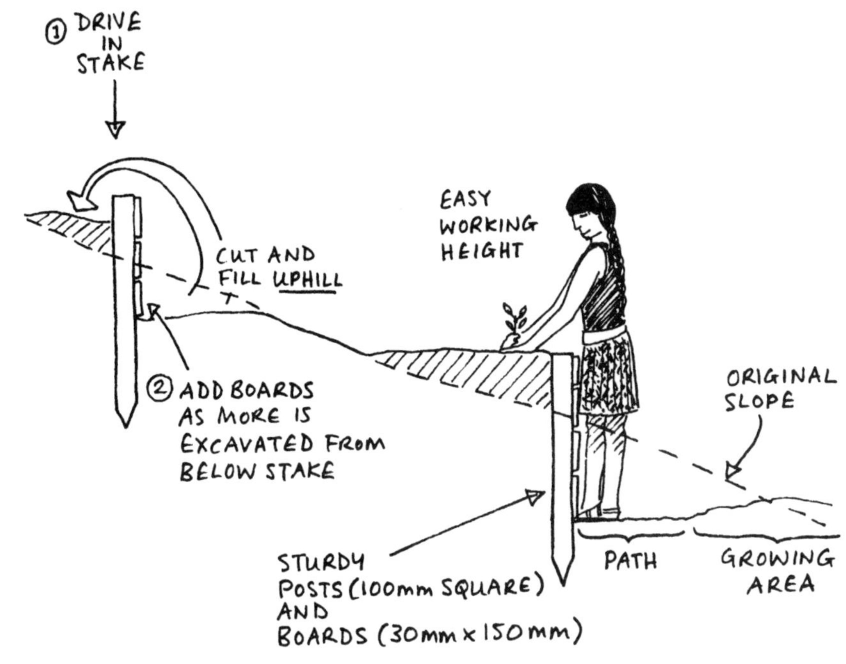

A low-input method for terracing a sloping garden.

arrangement should be strong enough to support the weight of the uphill bed.

Excavate the soil from below, always throwing it *uphill* behind the boards. When enough soil has been cleared more boards can be fixed to retain the lower part of the bank.

The process can be continued up and down the slope until it is all terraced. If you leave a path below each terrace and make each bed no wider than an arm's length you can garden uphill, and so avoid bending, whilst having made more no-dig beds on which you will never need to walk.

The level surfaces of terraces mean that cultivation can take place on even the steepest of banks. Continuous ground cover is important to prevent excessive evaporation in such free-draining conditions. Some crops prefer the dryness of a hill slope, such as grape vines.

There is a whole ecology of chalk uplands, for instance, which is composed of plants preferring open free-draining soils. Some of these crop heavily, so for instance the South Downs in Britain were widely planted with chicory during the First World War. This was used as horse fodder, thus freeing up valuable grain supplies for human consumption, and keeping grasslands available for meat and dairy production. Working to garden scale, this same approach may help keep all areas working for maximum benefit.

The vertical edges of terraces are appropriate (as we saw above) for rock garden cultivations or for hanging and climbing plants. Where aspect is appropriate they make good sun traps and heat stores for trained fruit trees and bushes.

In time, banking can be made more substantial by making stone walls. If well-planted, even slightly sloping earth walls may work–this is, after all, how all those paddy fields are made! Remember, in this and all things, the old principle: a little done well is far more likely to get done and succeed than an over-ambitious plan started with good intentions, but beyond your means.

seven

Adding Features

A house is a living-machine.

LE CORBUSIER, 1923

Every garden is an opportunity for rich variety in usage and in the interplay of edge and flat space. Additions to the garden can be tailored to the skills and space available.

Houses

For most people the major human construction in their garden is the home itself. Houses come in all shapes and varieties and a wide range of construction material. They may be very well placed in the landscape or part of some monumental urban development which is designed around maximum profitability from land area, rather than the prevalent weather conditions or seasonal advantages for a certain way of placing the house. It is probably true to say that in Western Europe and North America most people's houses just ended up where they are, rather than being consciously designed to be in their present situation.

We commonly talk of 'house and garden', as if they were two separate entities. In the sustainable garden we should assume that house and garden are part of an overall system in which one blends into the other. The house may itself be an area for growing food, from tender culinary herbs overwintered in the kitchen, through to mushrooms in the cellar.

Children can have great fun participating in indoor growing with small-scale ideas from growing sprouts in a sprout planter, available cheaply from most wholefood shops. A jam jar with a perforated lid can work just as well as the fancy bought versions. The old childrens' favourite of mustard and cress growing on the kitchen windowsill is a good example of producing fresh nutritious green stuff which is rich in vitamins, within the house.

Indoors is a good place to start new plants for the garden grown from pips and seeds taken from food consumed in the kitchen. It is also an opportunity to start off early seedlings which can then be planted out in the garden after the frosts are over. Unfortunately, however, seedlings started on windowsills can tend to get away too quickly, and suffer quite badly when transferred outside. There has to be a balance between enabling plants to get ahead early and making sure that they are hardy enough to withstand their transfer to the outdoor climate. Plants grown on windowsills should be turned often so that they do not become one sided in their growth through being attracted to the sunlight side.

Storage

The house also interacts with the garden in the way we carry and store materials from outdoors. We are always looking for efficiency in sustainable homes so that we can minimize labour. So it is important to have things like fuel stores carefully thought out. Do you store the logs for the fire where they're dry and handy on a windy winter's night? Or do you have to struggle like an extra from *Wuthering Heights* to a wet corner in the gloom? If you burn wood, then have an adequate small log storage space within the household so that on those cold, wet and snowy winter nights there is no tramping across the yard to get fuel.

It will also mean that the internal store gives a reasonable short term supply of well-dried wood which will take well on the stove. Where space allows, the system may evolve to a series of small storages for immediate usage, right next to the stove, with an intermediate storage perhaps in the back porch, and a main log store outside within easy walking distance.

The same sort of principle applies to gathering and storing food crops from the garden. It is great to grow an abundance of crops like apples or pumpkins, but next to useless if you don't have proper storage facilities. You don't want to walk a long way from the kitchen to get to the food store, but if you have grown 20 cases of apples in your garden over the summer you don't want to have to have all that fruit stored in your immediate kitchen area. One box at a time will do.

It is therefore inevitable that the house and garden evolve into a series of transition spaces. The porch is the most common example of these but many houses are entered through utility areas, boot stores, small corridors and so on which offer opportunities for storage, and in terms of plants themselves yield some things which will grow in these areas. Transition areas are also the way in and out of the dwelling and so it is important that they are as welcoming as possible. Careful thought about the design of a front or back door and the area immediately adjacent to it affects the whole feeling of the house for the residents as well as for visitors.

Making the transition from indoor tasks to working in the garden should not be a terrible effort; do you have to walk to the bottom of the garden, where

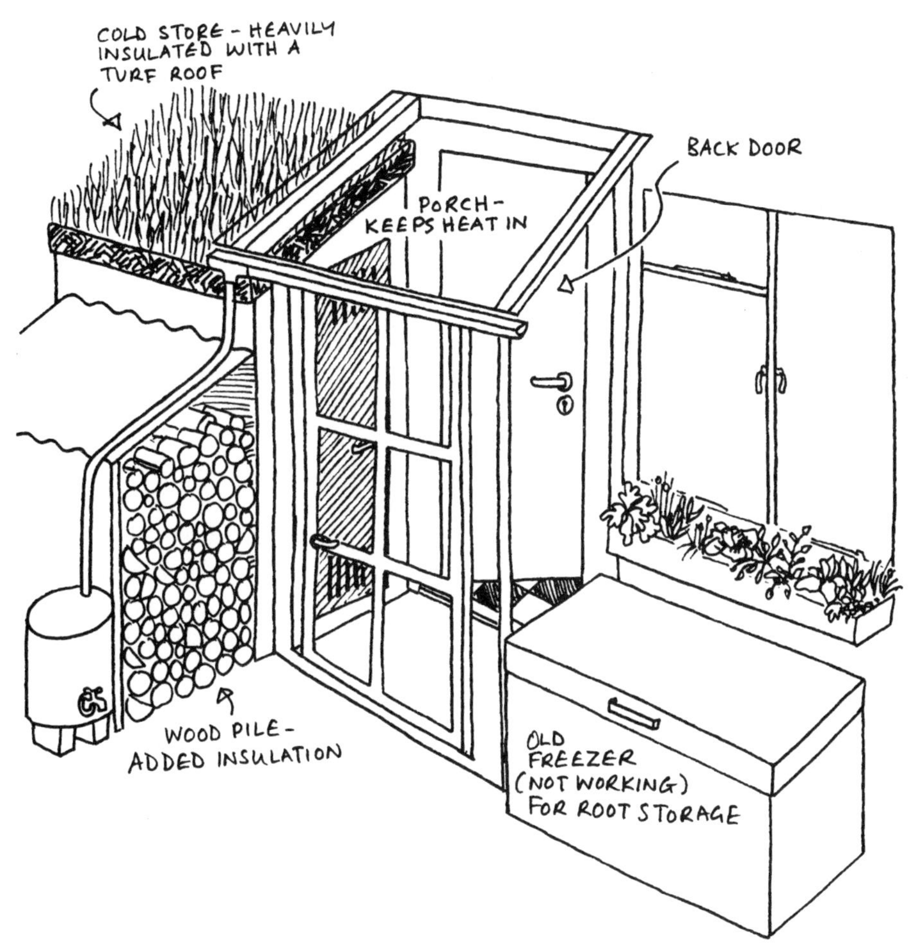

Easy access to suitable storage spaces can extend the times of the year when you enjoy garden produce.

the tool shed is 'hidden out of the way', to get a tool to do some work, or is there a handy storage place right by the back door? If it's the former you are disadvantaged. A bucket and trowel kept in the same place right by the entrance make it easy to pop out for five or ten minutes a couple of times a day to tend to the garden. This won't be possible if a big effort is required to get started.

Roof Areas

Buildings offer roof areas which are either flat (in building this term means a slope of less than 7 degrees), or sloping. An immediate advantage of these areas is that they enable us to collect water for use in the garden, and anyone whose household rain water drains straight away should try intercepting the down-pipes to collect water in rain butts in the garden.

Such areas are also growing opportunities. The traditional turf roof dwellings of rural Norway, for instance, offer not only a natural protection to the house, but additional insulation. Roofs can be covered in attractive plants such as herbs, which will give delightful scent on entering or leaving the building. You can be as inventive as you like in creating biomass roofing. All that is needed is an underlying impervious membrane to prevent water running through the roof into the building and causing damp and consequent dry rot. A living roof area can itself collect rain into gutters for water collection. The roof area so laid out can of course be a source of crops itself. Make sure that any collection from the roof area is safe, that is, made using properly secured ladders, and that any roof area walked on is strong enough to take the load.

Living Walls

The next areas to consider for yield in the household are the vertical areas of the walls. These are a natural opportunity for training plants either on or off the wall. On the wall training may require slightly less effort, and traditional systems use vine-eyes with wires running horizontally to give fixing points to the climbing plants.

Another option is to build up trellising on the wall of the house so that an air barrier is trapped behind the growing vertical plant material. This will also have the advantage that invasive plants like ivy will then cling to the

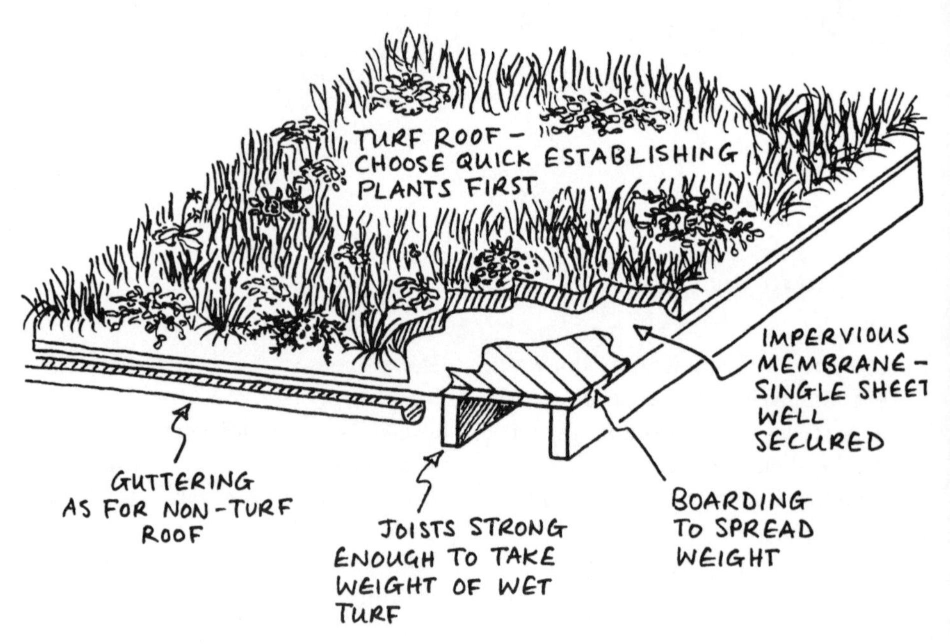

No opportunity for growing should be missed.

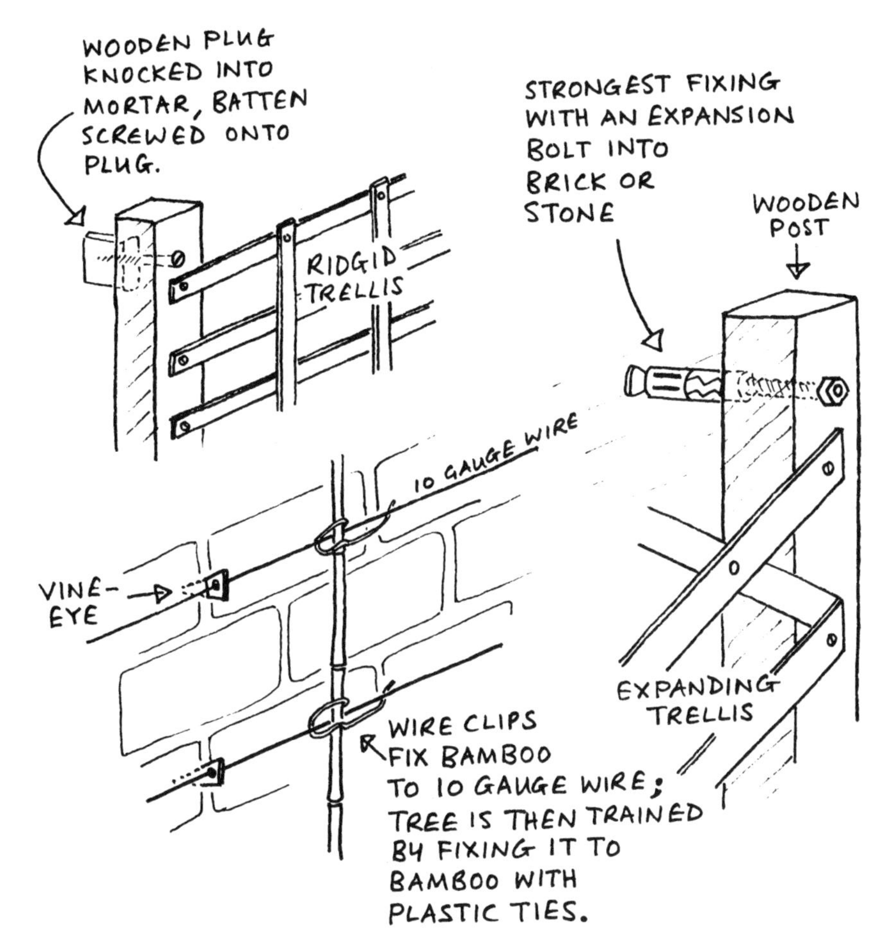

Careful fixing at the outset make for durable wall-trained plants.

trellising rather than to the house itself. This limits damage to the building through deterioration of the mortar between brick and stonework.

As with roof coverings, vertical plant growth against a house will offer insulation. Ivy leaves tend to lie vertically in the cooler months of the year when the sun is lower in the sky, offering a higher degree of insulation than in the summer months when the leaves lie in the horizontal plane venting the space between the trellis and the wall, and thus cooling the building.

House walls act as a solar store, trapping heat and thus making a warmer micro-climate in their immediate vicinity. This is particularly useful for ripening fruit, hence the tradition of training fruit up the sides of buildings. It is amazing that this is not done more often, particularly in smaller gardens where there is little space for full grown trees to mature without shading out other desirable crops.

Apples, pears, cherries and plums are easily trained in these circumstances. Where the climate is warm enough and the wall sheltered enough, other useful fruit crops are Kiwi fruits and grapes, apricots, peaches and figs. Blackberries and the various blackberry/raspberry-cross fruits will all prosper against a wall, although as they will also do well in a shady condition they might be better placed on shade-side surfaces and under trees, leaving sunny wall surfaces for fruits which need the sunlight rather more.

All these comments refer mostly to walls which are in full or partial sunshine. Walls on shady sides of the house can be used to build cooler food stores onto the end of the house, and as they are less useful for growing plants offer the best siting point for a recycling collection point. There are some climbers which will thrive in shade, such as winter jasmine, and *Hydrangea petiolaris*. If a back door vents onto a shady area like this, keep the bottles and the paper here before it goes to the municipal recycling area (or into your own ingenious designs).

Having said this, some fruit trees will do better in a shady area as this will encourage them to flower later in the year. Thus very early flowering varieties can be made to hang on until the more severe frosts are passed by the delayed development that a shady aspect will give them. For example, it is traditional to grow cherries on shade side walls in temperate climate areas.

The whole yield of the household garden transition area can be greatly increased by adding lean-to greenhouses on to the house. In many ways these have great energy advantages over free-standing greenhouses, as we shall see in the next section.

Greenhouses

The glass house is a marvellous invention. It causes minimal intervention while acting as a solar store: sunlight passing through glass is reflected and cannot fully escape again. The glasshouse itself becomes a store of energy from the sun, giving it a higher ambient temperature than the surrounding air. It additionally has the benefit that if attached to another building it reduces heat loss from the building, as air flows over the surface of that structure have been lessened.

Glass houses as entrances are good energy additions to households in another way: they reduce the volume of warm air lost from the house when entering and leaving the dwelling. A two-door system means that instead of losing large volumes of warm air from a hallway, you only lose (at worst) the volume of the porch or greenhouse. Provided, of course, that the household obeys the simple management system of always closing the door! Automatic door closers can help, but must be managed carefully to avoid catching young fingers on the return.

By trapping solar energy, glass

houses can also be used to warm potting sheds and adjacent chicken sheds or indeed any other structure which would be appropriate.

Glass houses can be extremely useful in extending the growing season or in allowing us to produce crops that prefer warmer climates than our own. Where I live in Southern Scotland in most years it would be foolish to predict a last frost earlier than the beginning of June, and we would expect frosts to have returned by September. This makes it difficult to get a good crop out of certain warmer climate species such as runner beans. By starting plants off in the pots in the greenhouses as much as a month or two can be added to the growing season for these crops. Not far from where I live there is a vinery which has been producing grapes commercially for 120 years, through the simple benefit of glass. We are talking the same latitude as Moscow in this case!

Greenhouses themselves offer the opportunity for rainwater collection and I have seen very good designs where the downpipes and the guttering transmit the rainfall to holding tanks inside the greenhouse, making minimal work for watering the greenhouse crops. If placing much reliance on greenhouses as a source of food it is good to think of making them self-watering and thereby reducing the

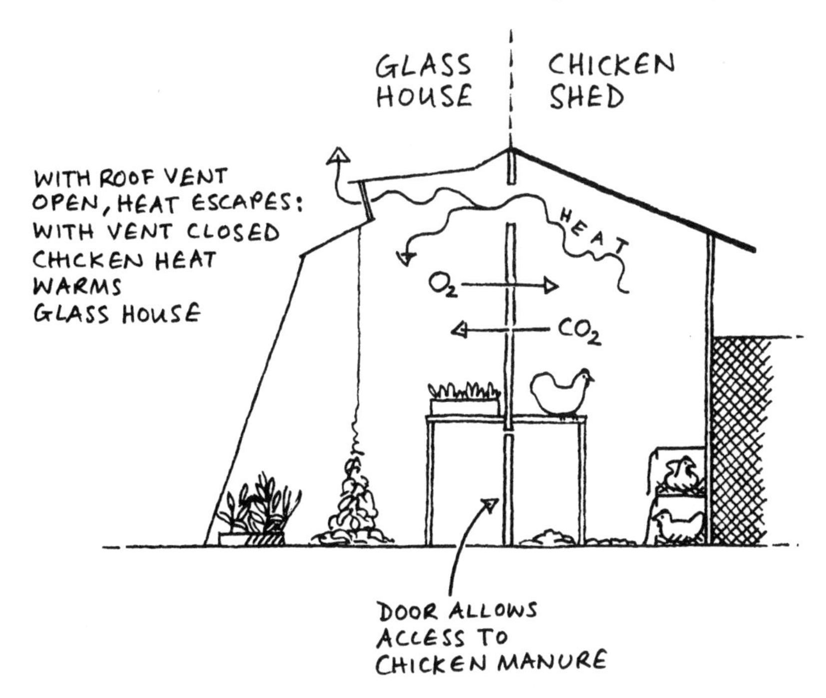

The chicken greenhouse is a splendid idea for redirecting energy to where it is useful, instead of wasting it!

amount of labour they cause.

Attention should be paid to the need to control the micro-climate of the greenhouse through the seasons. A favourite Victorian example of this was to plant a grape vine outside the greenhouse and grow it through a hole in the wall so that no watering of the plant was necessary, its roots being out in the full benefit of the rain. The plant itself produced leaves and fruited within the greenhouse, with the advantage that, as a deciduous plant, the greenhouse was kept cooler in the summer by the shading action of the vine leaves, but admitted maximum sunlight in winter when the vines were bare.

There are sophisticated devices, such as thermostatically operated vents, available to cool and shade greenhouses and prevent them getting too hot in the mid-summer heat. These have the advantage that they are not dependent on you remembering to crank open the vents to prevent all the plants burning up.

It is also helpful to think of relatively 'passive' ways of warming greenhouses. This is particularly important in the beginning and the end of the summer when the greenhouse can get quite cool at night. One way of doing this is to have a grey water pool for waste water from the kitchen and bath which is ponded within the greenhouse before draining elsewhere in the garden. This gives an intermediate store which allows all the heat still remaining in the used washing water to escape into the greenhouse and help warm it. Another way of warming greenhouses biologically and effortlessly is by keeping chickens in them overnight. If your home has been designed to have such energy inefficient measures as fire places backing onto external walls, this again is a good place to build greenhouses outside to trap the heat which would otherwise escape through the exterior surface of the house. Compost heaps within a greenhouse will also provide some extra heat.

There are many inviting designs for 'conservatories' these days which suit the deeper pocket. It is important to remember that a greenhouse does not have to be a wonder of aluminium and expensive glazing to be effective. Many good greenhouses have been built from scrap. Tony Wrench had the marvellous idea of making greenhouses from scrap car windshields:

> *Hold a windscreen in your hands–*
> *heavy, smooth, so optically perfect–*
> *and you become aware that to any age*
> *or culture but our own, this artifact*
> *would be a marvel beyond price.*
>
> TONY WRENCH, *PERMACULTURE NEWS* #23, 1991

Available as scrap from car and truck breakers, you build them up on wooden frames, overlapping enough to let rain run off.

As more and more people modernize old buildings and take out old windows and throw them away these become a very good source of recyclable material for building cold frames and greenhouses. Simple and cheap greenhouses can also be made from clear corrugated plastic. For those who can make a larger investment, such modern materials as triple glazed polycarbonate offer energy-efficient material. At Achiltibuie in Wester Ross, Scotland, at latitude 58 degrees N they are able to grow bananas in unheated greenhouse areas. The careful angling of the glazing material to the solar aspect has measurable effects on the climate within the greenhouse, and even at latitudes which are sub-arborial (that's next to tundra on the scale of inhospitality) it is possible to maintain a sub-tropical climate within their structure.

Tony Wrench tried another innovative idea for a quick and easy greenhouse recently in Wales, weaving a dome from four-metre-long willow cuttings. He covered the dome in heavy duty UV polythene (stocked by agricultural merchants), and stressed the importance of making the cover from a single sheet, taping down any folds.

Imagination is the chief material in these designs to extend the garden output.

Mother Earth and Father Sun willing, the supports will sprout leaves in a month or so. In August, plenty of leaves will supply shade from the mid-day sun, but there will still be ample light. In November, Autumn will arrive late inside the dome; the leaves will fall to allow more light until next year's early Spring. After four or five years, when the soil is a bit tired, I will remove the plastic and have a magical bower inside a growing willow coppice.

TONY WRENCH, *PERMACULTURE NEWS* #15, 1989.

(This is a variant on the bower design shown in Chapter 1, p. **15**.)

One of the things to remember with greenhouses is that because the soil is not exposed to the external environment it can become rapidly soured, and so prone to pests. If using greenhouses on any broad scale within the garden one solution to this is to build a structure on runners, so that it can be moved over fresh ground each season. It is only necessary to have a run of one or two pieces of ground and to work last year's plot with a different crop, such as beans, for regeneration of soil before it becomes an internal growing area again.

Otherwise it will be necessary to dig out and renew greenhouse soil from time to time. There are some good ideas for biological husbandry for keeping down pests in greenhouses, including the idea of cleaning them out periodically with fowl such as ducks, geese or chickens. An occasional sewing of green manure will be helpful.

Trellising

The smaller the garden the more important it is to increase the vertical growing area. It is possible to double the growing area in a garden by the simple use of trellises, to create a growing area for fruit, pole beans or other climbers. In addition, trellises have the advantages of offering more opportunities for micro-climate by creating shade spots and sun traps.

Vertical areas can be used to accommodate both true climbing plants, and various plants which can be trained to grow upright, although this is not necessarily their normal habit. So honeysuckle is a true climber, but various shrubs such as *Chaenomeles japonica* (ornamental quince) can be trained to grow upright by attaching them to a vertical structure.

Trellises can go from the straightforward fence type into very imaginative structures offering such superb features as shaded walks and pergolas. Areas shaded by living plants provide welcome relief from a hot summer's day and can add to the living area of the house and garden by offering attractive places for working or eating out of doors. We have already mentioned the need with young children in the family to have shady places to play in the garden so that children are not subjected to the unwelcome heat of the full summer sun.

Trellises do not need to be constructed from expensive bought components, although a lot of these sort of materials look very attractive. There

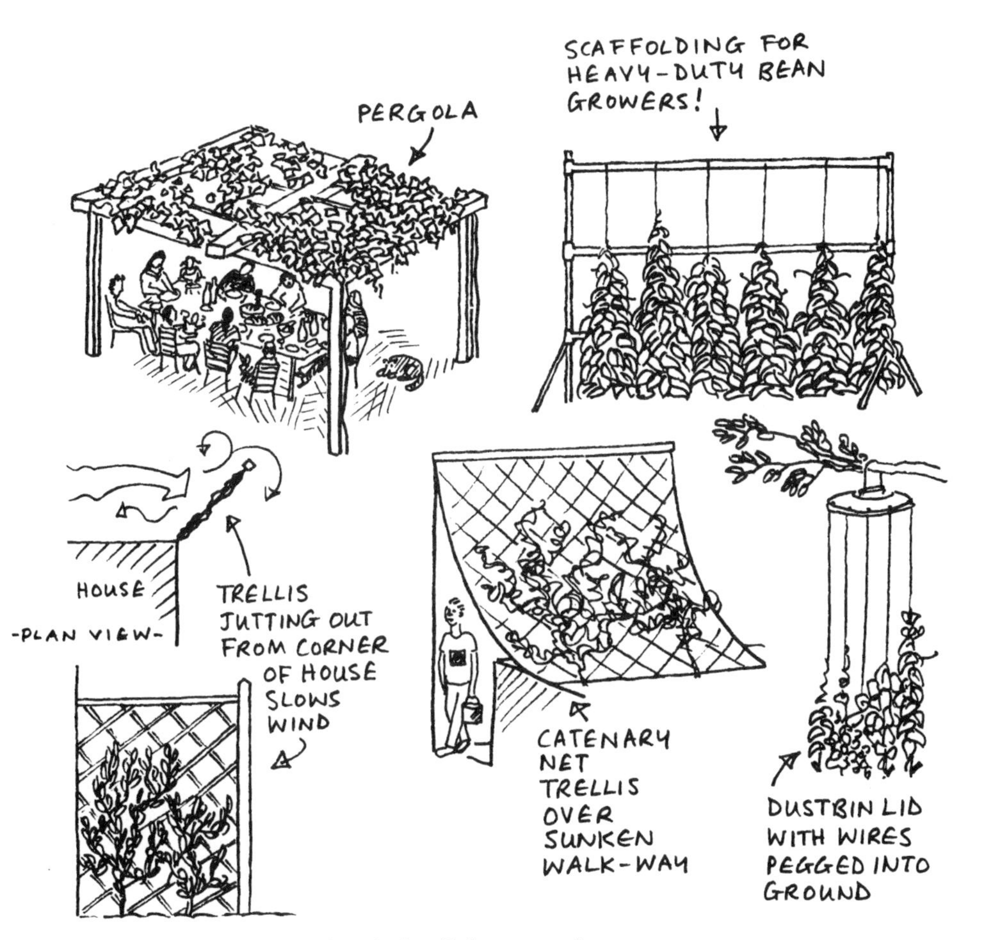

The air is all free growing space.

is plenty of opportunity for building with scrap in the garden and with care and attention to detail such structures can also look very attractive. When building with wood in the garden it is as well to be aware of which timbers last well and to ensure that the wood used is preserved as well as possible. Modern preservatives give an even finish which blends wood colours together. A bit of careful sawing and brushwork can make scrap wood look as good as new. Do be careful to pull out unwanted nails though!

Many people have moved away from the old favourite of creosote, to more environmentally acceptable wood treatments. However there are some advantages to creosote as a home made way of using up old sump oil from car engines. In many cities this is just poured down the drains after people have done their home car maintenance. By saving up old oil in large metal or heavy duty plastic barrels (carefully inspected to make sure they are leak free), timber for use in the garden can be left to soak in these barrels, and taken out allowing any surplus oil to drain back into the barrel.

Another way of preserving wood from rotting when hammered into the ground is to char it in a hot fire. The charred external surface is much more resistant to rot than the natural timber.

Trellises can also be constructed of wire slung between posts, and these can be designed as successions so that plants trained along the wires will in time become living structures themselves, rendering the wire no longer necessary. Indeed it is possible to grow trellises which are entirely living structures. One example of this would be to plant a tree and a vine into the same hole, and to cut the tree back at two to three years old to make a horizontal espalier, along which the vine was trained. The whole evolution of this sort of system brings us back to the idea of hedging in the garden as a multi-functional construction.

Access: Paths and Roads

The main consideration with building paths, or if your garden is big enough, roads, is to make them as much as possible from materials that are locally available. Pathways should be sufficiently hard wearing to stand the traffic which they are designed to take, and it is helpful if they are resistant to rain and snow. It also helps if they drain properly so that they don't become quagmires in wet weather. Ideally they should have a slightly sloping surface area so that they drain freely, and paths are themselves a very good system of drains if well designed. Any water so caught should be redirected back into useful places in the garden, so that all access ways become part of the self-managing water collection system.

Any sort of crushed chippings or pebbles are useful for making pathways, provided that the surface has been underlaid with some relatively impervious material to prevent deep rooting weeds from becoming a problem. Turf paths are relatively easy to maintain. Although they take time to mow, that is the only treatment required to keep them in good shape. Many people have set out to make gravelled paths and then regretted the amount of time that they have had to spend in weeding them to stop them reverting to high forests and raking them to redistribute the gravel.

If your growing areas are not all reachable without walking on them it is useful to have a supply of temporary pathways. This might be something as simple as planks that can be laid down between rows, so that when you walk on them your weight is spread and compaction of the soil is minimized. Another nice idea for temporary pathways is to make a roll-out bundle from slats attached to ropes on each end; these give you a walkway which is easily moved from place to place and can be stored in a small area.

Particular attention needs to be given

to pathways on slopes. Here it is even more important to make a surface that is resistant to erosion, but which also affords safe footholds in wet conditions. Spare kerb stones are often available, particularly where old pavements have been taken up. These provide a very hard wearing and attractive front step, the back of which can be filled in with gravel to make a firm footing.

Stout planks can be affixed as step ways with stakes in front to prevent them coming downhill. Wooden steps can get very slippery in the winter months and a small area of fine chicken wire stapled to the top of them will prevent the inevitable tumble which comes with an insecure foothold. If making the garden accessible for users with wheelchairs and prams, then it is important to make pathway surfaces that are of a low enough gradient to be easily passable, as well as being wide enough for wheelchair and pedestrian to pass each other.

There are some useful plants which like to grow on pathways and in fact can greatly enhance their nature. Many of the smaller succulents such as sedums are good for this purpose. Some plants such as lawn chamomile actively thrive on pathways, and have the additional advantage of yielding very pleasant smells when walked on.

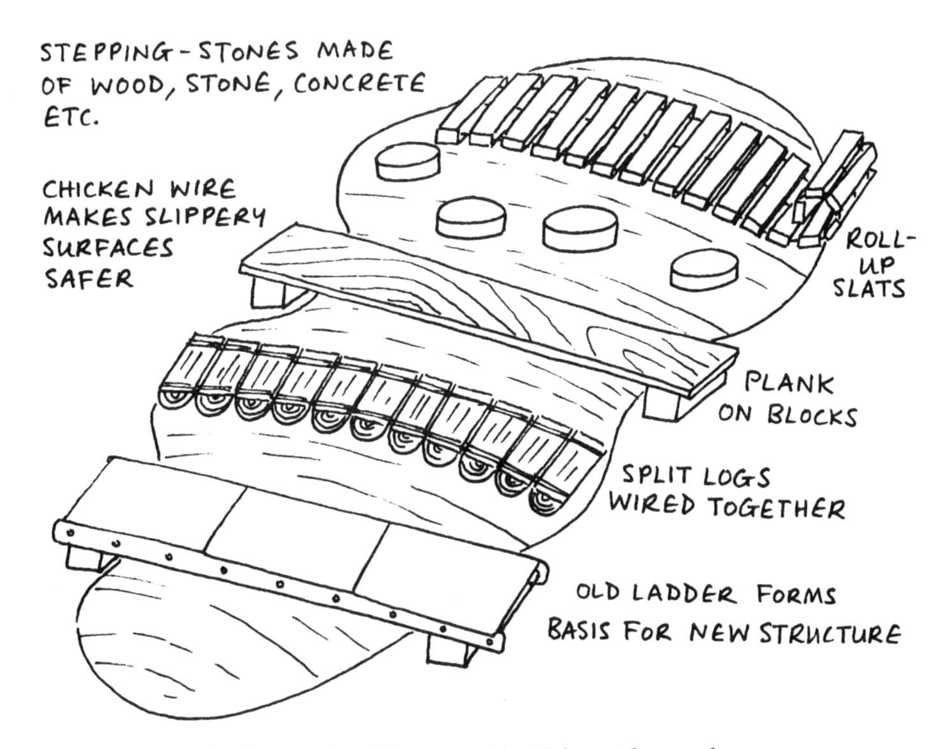

Paths need not be space 'lost' from the garden.

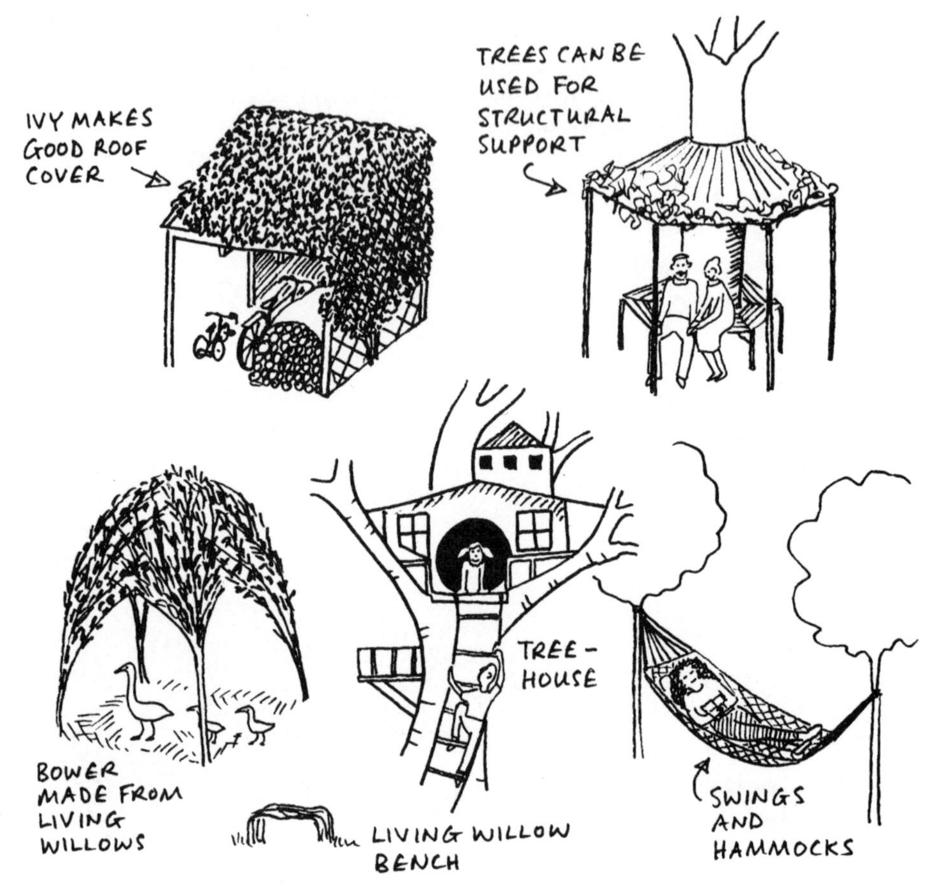

In the garden buildings come alive.

Another idea for a hard-wearing surface which gets away from the concreted look is to lay open-slatted bricks across the ground. The pockets in these are filled up with soil and seeded. This gives a surface which is hard enough to withstand motor vehicle traffic but from anything other than close-to looks like green space.

Attention needs to be given in a garden to properly fencing and providing gates. Think of gates as barriers which can be easily closed and which serve their purpose in terms of either keeping children in or keeping unwanted people and animals out, but also which can be easily opened if you are trying to get through them with arms full of shopping, or a wheelbarrow full of good stuff for the garden. There are some attractive ideas for self-closing gates, and the B.T.C.V. Handbooks (see Booklist) are full of great designs to suit different rural styles.

Washing ~

Your average garden book tends to forget that for most of us our backyard is where we perform lots of domestic functions. One of the most common of these is hanging out the washing to dry. Does the washing spoil our garden by giving it lots of untidy flapping shirts and undies, or can we see it as a positive advantage? I have often wondered about the potential of washing as a bird scarer in the *Brassica* plot.

Washing either needs to be hung over an area where we can walk, or it needs to be put onto lines which can be retrieved from an area that we don't wish to cross. It is also worth thinking

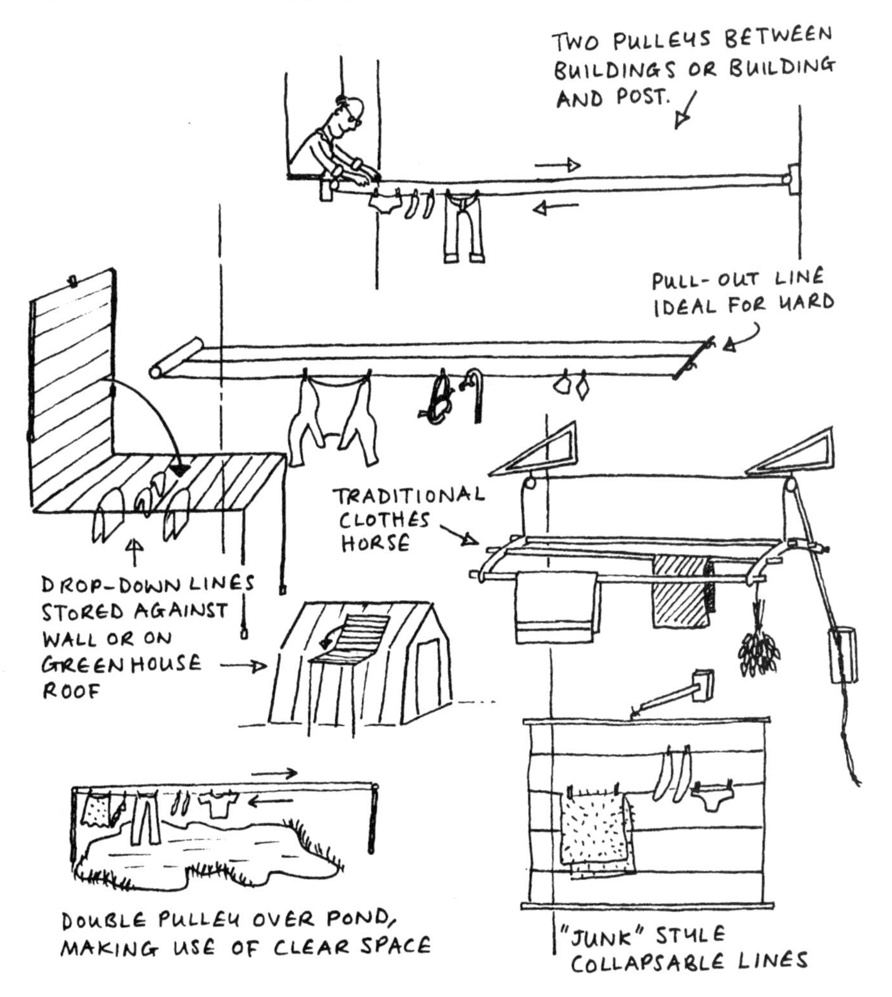

Here are a few designs for some line and pulley systems which set out to do little more than simply get the washing out.

of the effect of water draining off dripping washing as having some benefit to plants underneath it.

Remember that the washing should be hung close to the house so that it is easily collected in again in the case of that sudden shower that seeks to undo the work of a good morning's sunshine.

The bleaching power of solar rays is very beneficial in keeping our clothes feeling fresh and clean, and an outdoor airing for our precious garments is as good a use for the garden as we are likely to see.

Play Space

Likewise for most families the garden is precious as an area for children to play. This partly depends upon the age of children who are likely to use the play area. Younger kids will be less inclined to need climbing space, and flat areas for older kids might need to be bigger to allow for more rumbustious games.

Flat areas should be provided in any case for whatever needs the family has, from baby crawling on a blanket, to a full-scale football bout, to a family

Ingenuity, timber, nails and a spare weekend can produce challenges to everyone's play ability.

picnic. The traditional lawn can be extended into a more interesting growing area of mixed herbs and edible plants so that it offers yield as well as the inevitable work of the mower. Biological mowers (rabbits, geese or sheep) are another way of turning a problem into an asset.

Wood chips make another hard-wearing play surface which is relatively safe when children fall over. Climbing areas might be integrated with trees or trellising, although obviously you will need to make sure that the constructions are stout enough to take the weight of whoever is going to be playing on them. Also be careful that, if making climbing areas, surfaces below are soft enough to receive falling children kindly.

Children will delight in sandpits and water in any garden. Water needs to be managed to make sure that it provides no risk of drowning for smaller children: remember that children can drown in as little as two or three inches of water and that, if the water so provided is solely for playing, it should be easily drainable after use with supervision.

There is no reason why children's play equipment can't itself do useful work; the designs above provide some creative ideas. Many of these constructs and others which you yourselves will dream up along the same lines can easily be made from scrap timber. Be very careful that the timber is splinter free, and that any old nails or screws or bolts have been removed from them before using them in the garden.

Then you'll need more tools and extra materials. Look for second-hand heavy duty timber, 15 x 7.5cm (6 x 3 in) upwards, used car and tractor tyres, strong exterior-grade board, lengths of old chain (5cm [2in] links), big nails and bolts. You can get as fancy as you like, but don't leave anything for kids to play on that isn't *completely* childproof and safe. If children can hurt themselves on anything, they will.

The tools needed for garden timber work are fairly simple: a hammer and a saw, as well as a drill and spanners if you're bolting timbers together. If using an electric drill outdoors (or an electric mower) make sure you have the right electrical equipment, that is:

- an adequately large extension lead (13amp in the UK)
- a residual circuit current breaker (RCCB) which can be bought as a plug-in adaptor if your household supply isn't fitted with one.

DO NOT use electrical equipment outdoors without proper safety devices.

Zones

Structures can be used very usefully in the garden to break down different zones of activity. Too many gardens are designed as small rectangular shapes with very unexciting aspects offering minimal use of edge. By using trellises, play equipment, sheds, greenhouses, walls, hedges and any other of the ideas contained in this chapter a much more vibrant space can be created. Every structure has the possibility to have more than one use. By using structures as the boundaries between zones it is possible to take advantage of this way of thinking.

eight

Water in the Garden

Water is excellent.
PINDAR (518–438 BC)

In the struggle for achievement it's easy to forget the simple things. Without water there is no life. We should pay it maximum attention.

Where it Comes From

Water is a precondition for the presence of life on the planet earth. Living organisms use water as the medium by which they carry sustenance around their bodies. This is equally true of plants and animals. Most water in the garden is invisible to us because it is working away inside growing plants.

Of course there are also opportunities for using water in the garden visibly. Ponds and streams can be attractive and productive not just in themselves, but in the wide range of life they support.

To be useful water needs to be of adequate quality and quantity. This means that water used in the garden needs to be reasonably oxygenated, free of toxic material, and 'fresh'. By fresh water we normally mean water which is very low in salts in solution. At 1 per cent salts in solution water is termed brackish and 3.5 per cent constitutes salt water. Apart from plants and other creatures specifically adapted to living in sea-water environments, this water would be useless in a garden. There is always the extra challenge of designing a garden for a seashore environment based on plants that thrive in these conditions!

In the developed world where water is coming to be regarded as the precious commodity which it truly is. Many urban civilizations are now struggling to source themselves adequately with good water, and the consciousness of the need to preserve supplies is becoming as high in the developed world as it has always been in the developing world.

Table 16 Salt Tolerant Plants

Plants with high salt tolerance for coastal gardens.

Sycamore	*Acer pseudoplatanus*	P Hedging
Seabeach sandwort	*Arenaria peploides*	P Erosion resistance
Sea beet	*Beta vulgaris*	B Green vegetable
Hottentot fig	*Caprobrotus edulis*	P Edible
Scurvy-grass	*Colcheria officinalis*	P Edible
Sea kale	*Crambe maritima*	P Cabbage substitute
Hawthorn	*Crataegus* spp	P Fruit and edible leaves
Rock samphire	*Crithmum maritimum*	P Delicious edible
Broom	*Cytisus scoparius*	P Nitrogen fixer
Sea holly	*Eryngium maritimum*	P Edible
Sea milkwort	*Glaux maritima*	P Pickles
Sea purslane	*Halimione portulacoides*	P Edible
Sea buckthorn	*Hippophae rhamnoides*	P Edible
Iceplant	*Mesembryanthemum crystalinum*	A Edible
Pine	*Pinus* spp	P Hedging
Sea plantain	*Plantago maritima*	P Erosion resistance
Blackthorn	*Prunus spinosa*	P Hedging/edible fruit
Marsh samphire	*Salicornia* spp	P Edible
Goat willow	*Salix caprea*	P Hedging
Stonecrop	*Sedum album*	P Edible
Milk thistle	*Silybum marianum*	B Salad/herb
Alexanders	*Smyrnium olustratum*	P Edible herb
Whitebeam	*Sorbus aria*	P Firewood crop/berries
Swedish whitebeam	*Sorbus intermedia*	P Firewood crop/berries
Tamarisk	*Tamarix gallica*	P Hedging
Holm Oak	*Quercus ilex*	P Hedging/nuts

In general the water cycles of the planet are driven by an enormous nuclear reactor, known more popularly as the sun. This is responsible for evaporation from the seas, which forms the greater part of rainfall which falls on land. It also creates currents in the air which cause the winds and therefore create our patterns of climate.

Precipitation has many forms, the most common of which is rain. However, particularly in latitudes nearer the poles, and in mountainous and desert regions, a significant proportion of precipitation may arise during the year in the form of snow, frost or dew.

It is a good idea to know the rainfall for your area, and the times of year when it is high and low. Heavy soils

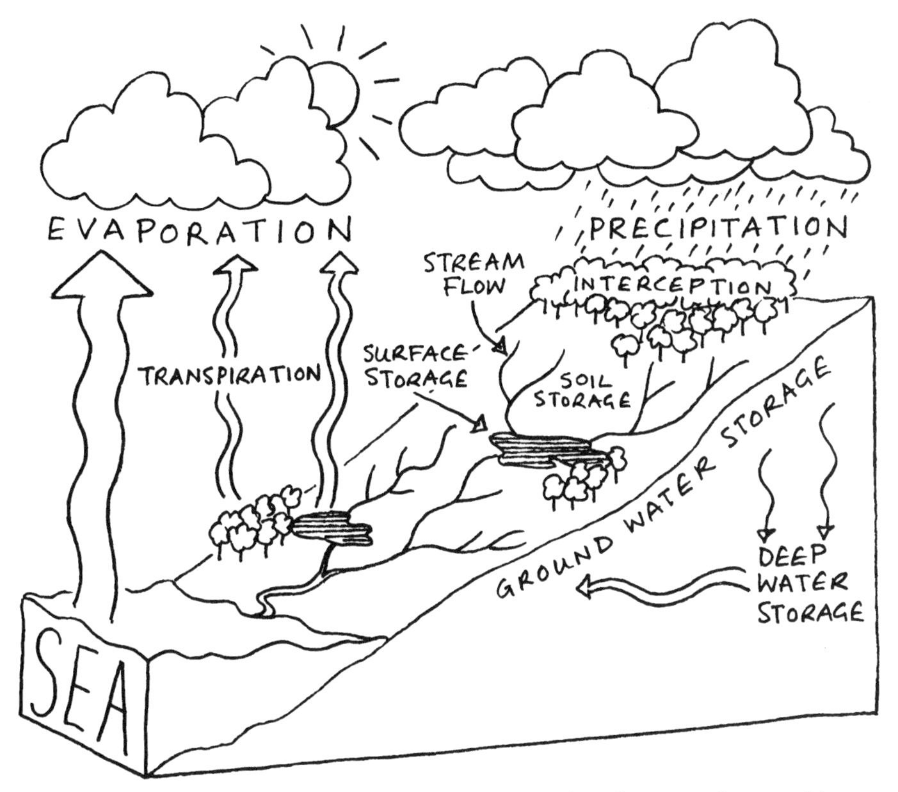

Water comes and goes from the garden in cycles that are planet wide.

should not be worked when wet, as this leads to compaction, so preparation of beds should be scheduled for drier spells. Even in a small country like Britain the rainfall can vary by a factor of five from place to place.

In the northern hemisphere there is a tendency for Western locations to be wetter as they are closer to maritime origins of the prevailing winds. In the southern hemisphere this tends to be reversed. Beware, however, that these are broad scale generalizations. The micro-climate can vary enormously in very short distances.

The quantity of available rain will tell you the extent to which you need to protect against damp, or allow for drought. There is not much you can do to increase precipitation in your garden, but you can increase collection and storage to allow for dry conditions.

There is even less the gardener can do about the *quality* of water in rain. If living in an area of acid rain caused by wind-borne industrial pollution, the best that the gardener can hope for is to lime the ground and any standing water to counteract the effects of acidity. This may be particularly important in urban areas.

A management solution to deal with

the effects of poor-quality rain would be to maintain a balanced mixture of plants in the garden, including a good range of perennial species, plenty of herbs and green manure, as this combination will tend to give a good mineral balance to the soil. This will be enhanced if all crop wastes are returned to the soil. Careful application of this principle will tend to counteract the effects of airborne pollution.

Storing It ~

Water quality is maintained by careful storage. Water can either be stored in flowing conditions, or in still tanks. You will generally only be able to take advantage of flowing water if a stream happens to run through your plot already. In very large gardens, where there is enough gradient in the land, and enough sources of water, it may be possible to create new streams. The number of people in this lucky situation must be very small. If you are, then supply can be increased (and made more sustainable) by pumping water from the bottom back to the top.

Solar power, wind power and ram pumps will all be able to do the job to some degree. Electrical pumps will be more reliable, but not particularly sustainable as regards their energy source.

For most people static storage is the best they can hope for. There are then two options. Will the water be stored in the open, or under cover? Let's look at each option in turn:

Open Storage

If keeping water in ponds, there are a number of risks. Evaporation reduces the available water store. This can be reduced by keeping the ponds relatively small in surface area; many small ponds are better than a large lake. The deeper the pond compared to surface area, the smaller losses will be.

Shading the pond with overhanging trees and shrubbery will also reduce direct sunlight, and therefore the rate of evaporation, as will plants with leaves on the surface of the water, such as water lilies. This strategy will also help protect against the second risk, which is decreasing quality of water.

It is natural for standing water to be surrounded by vegetation. This is an important part of the life cycles of aquatic areas. Leaffall feeds the bottom of the pond with decaying plant matter. This feeds bottom rooted plants, and to some extent bottom feeding fish. The trees also provide a habitat for insects which, in falling to the surface of the pond, provide for surface feeding fish.

A vigorous population of living creatures is the best indicator the gardener could have that the water quality is satisfactory. Insects will arrive unbidden; fish must be chosen and introduced. Fortunately there is a growing number of specialist stockists who can advise on balanced populations of livestock for those wanting to start a pond.

Another risk to the pond life is that of pollution. Ponds must be sited to minimize risk from polluted water run-off (such as road drains), spray drift (from non-organic farms and gardens), and

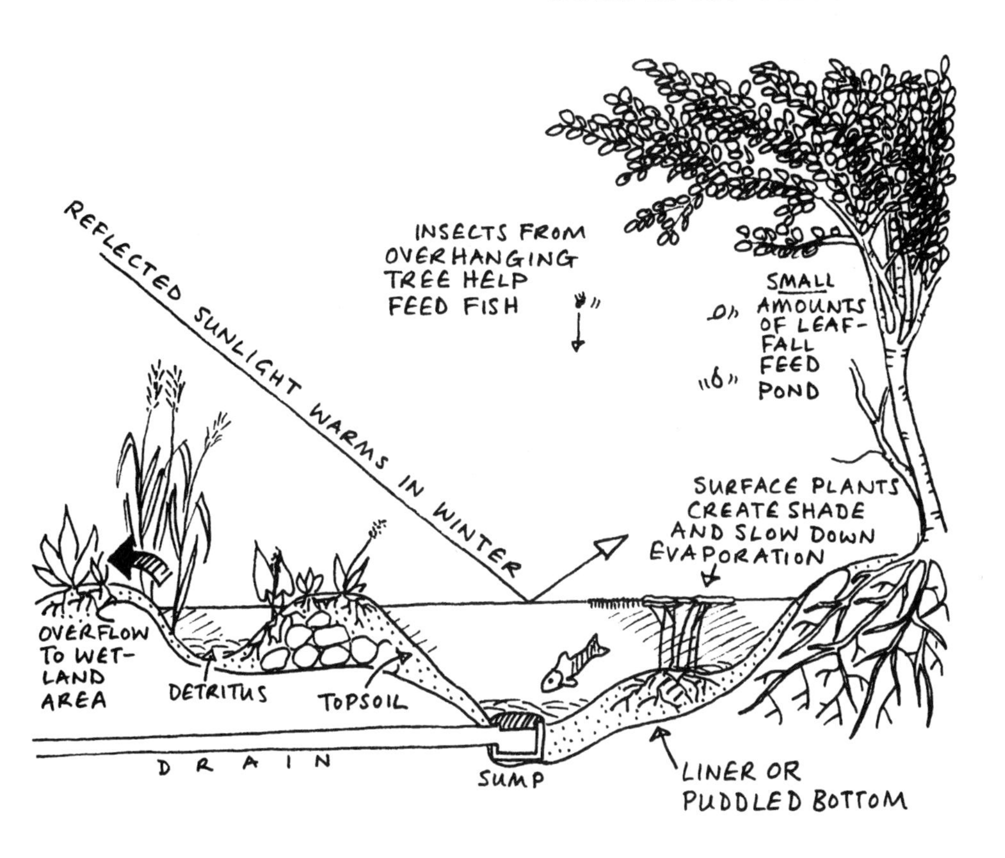

Your garden may not be able to accommodate a pond with all these features—but the principles can be applied anywhere.

general airborne pollution. Netting to protect from fallen leaves in the autumn will be helpful: a sudden glut of leaf material can kill the pond. There must also be provision in the design of the pond for future cleaning operations.

Ideally it should be possible to drain it easily, to clear out any build up of bottom sediment. It should be possible to net the surface of the pond to remove wind-blown debris. In some commercial fish farms it is common to see dead fish left on the bottom of the tanks; obviously this is poor practice, not to be recommended.

If the pond is going to provide irrigation water, then it needs to be designed with an upper and lower limit. Overflow should be used to make a wetland garden, favouring water margin plants. Irrigation of other areas of the garden will be easiest if the pond occupies a high point in the garden, allowing passive flow.

Table 17 Water Plants

Plants that favour a damp or wet environment.

Common name	Botanical name		Habitat
Sweet flag	*Acorus calamus*	P	Water edge
Marsh mallow	*Althea officinalis*	P	Wetland, salt marsh
Flowering rush	*Butomaus umbellatus*	P	Wet meadow
Marsh marigold	*Caltha palustris*	P	Water edge/surface
Earth almond	*Cyperus esculentus*	P	Water edge
Water hyacinth	*Eichornia cressidus*	P	Water surface
Water chestnut	*Eleocharis dulcis*	P	Water bottom plant
Hemp agrimony	*Eupatorium cannabinum*	P	Wet banks
Joe Pye weed (US)	*Eupatorium purpureum*	P	Wet banks
Meadow sweet	*Filipendula ulmaria*	P	Wet meadow
Water avens	*Geum rivale*	P	Wet banks
Yellow flag	*Iris pseudacorus*	P	Wet banks
Rushes	*Juncus* spp	P	Wet meadow—water edge
Duckweed	*Lemna* spp	P	Water surface
Lovage	*Levisticum officinale*	P	Wet banks
Gay feather	*Liatrus spicata*	P	Wet banks
Gipsywort	*Lycopus europaeus*	P	Wet banks
Purple loosetrife	*Lythrum salicaria*	P	Wet banks
Bogbean	*Menyanthes trifoliata*	P	Acid bog
Bog myrtle	*Myrica gale*	P	Acid bog
Water mint	*Mentha aquatica*	P	Water edge
Water cress	*Nasturtium officinale*	P	Water edge/surface
Water lily	*Nymphae tuberosa*	P	Water bottom/surface
Common reed	*Phragmites communis*	P	Shallow water
Bistort	*Polygonum bistorta*	P	Wet meadow/water edge
Fleabane	*Pulicaria dysenteria*	P	Wet meadow
Lesser celandine	*Ranunculus ficaria*	P	Wet banks
Giant rhubarb	*Rheum alexandrae*	P	Wet banks
Bulrush	*Scirpus* spp	P	Water edge
Sphagnum moss	*Sphagnum plumilosum*	P	Wet woodland/acid bog
Soapwort	*Saponaria officinalis*	P	Wet banks
Valerian	*Valeriana officinalis*	P	Wet banks
Brooklime	*Veronica beccabunga*	P	Water edge/surface
Duckweed	*Wolfia* spp	P	Water surface
Wild rice	*Zizania aquatica*	A	Shallow water

A secondary use of ponds is to increase available sunlight by surface reflection. Site the pond so that its shade-side boundary can benefit from this, especially in winter.

Closed Storage

It is less work initially, and certainly less work in maintenance to use closed storage. Well-sealed tanks admit no rubbish and are virtually immune to evaporation. However, water quality can suffer because it is a comparatively lifeless environment.

It also makes it more crucial that water quality is good enough at the moment of collection, for there are no living processes to cleanse the supply. Filtering through charcoal is a good method for cleansing collected water. In the tank itself alkaline material will help keep the water sweet. Crushed chalk or limestone is ideal. Closed water storages can also be used as heat stores over fairly long periods of time if properly insulated. In this way summer sunshine can be stored in your water tank to become background heating in

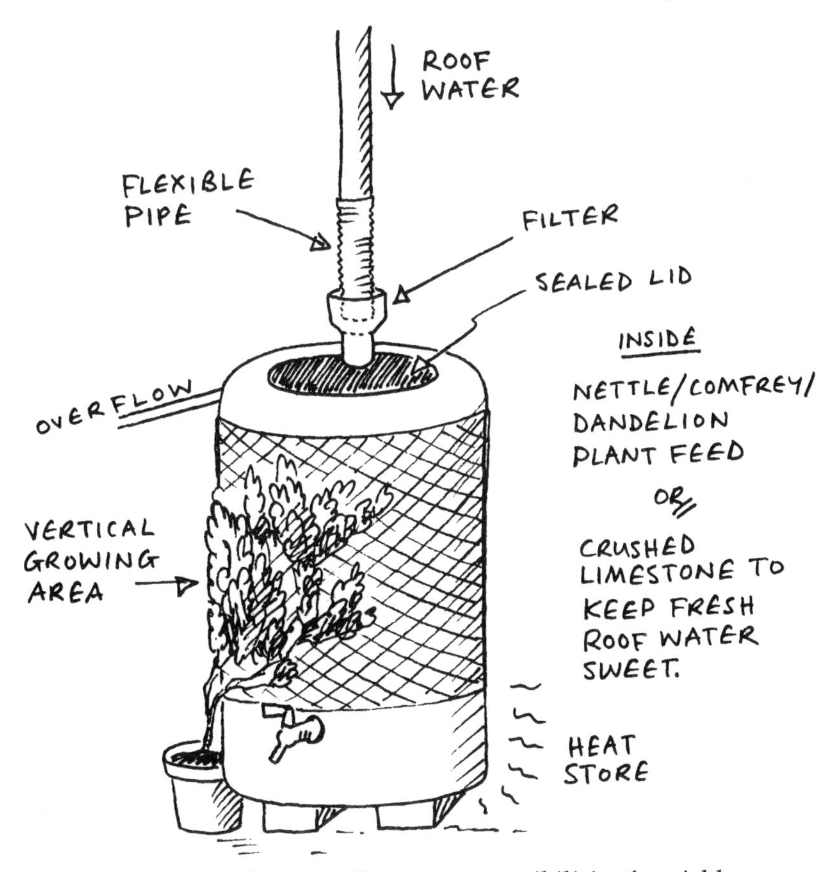

Even a tank of water offers many possibilities for yield.

winter. This will need careful design and must be integrated into the building structure to work. (See *Living with Energy*, in the Booklist, for examples.)

Closed tanks are also opportunities to turn water into concentrated plant feed. Nettles, dandelions and comfrey all add minerals and nitrogen to the water. Whilst the output can be black and smelly, it is certainly very healthy for the plants.

On a less ambitious level note that, for watering plants, tap water is often far cooler than the plants would like. Outside water butts have the advantage of bringing the water to the ambient temperature, and making it more acceptable to plant needs.

Collecting It

As we saw when discussing buildings in the landscape, we can use the surface area of the house and any sheds in the garden as collection points for concentrating the supply of rain water.

In addition there is the opportunity of recycling water from the domestic usage of the household. Waste water

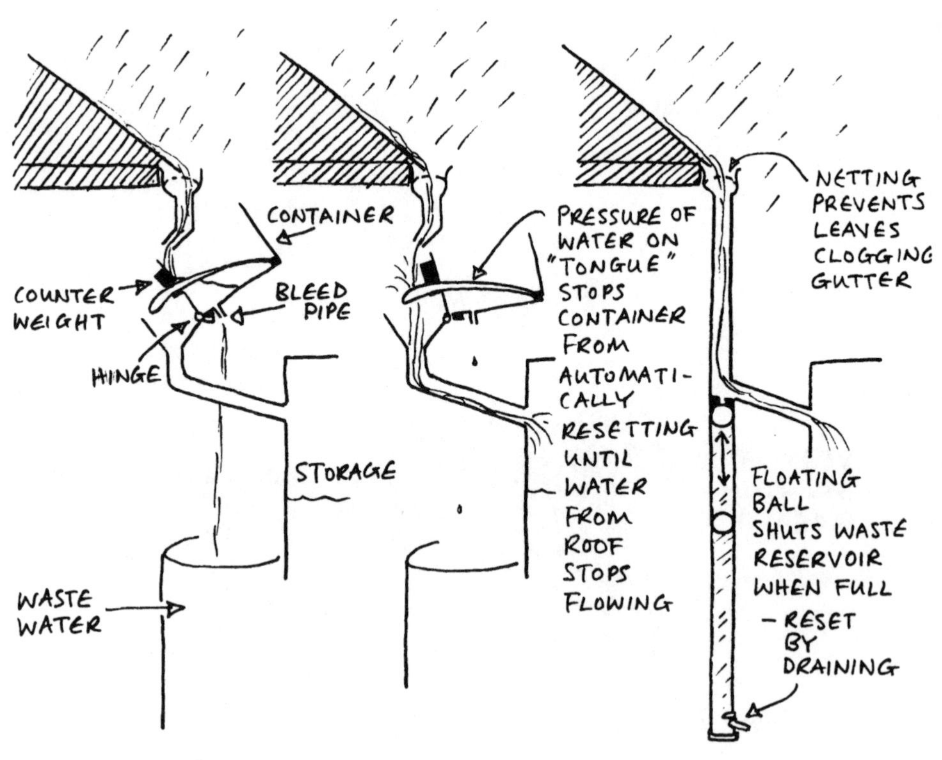

A little research and a few practical skills can bring self-sufficiency in water nearer reality.

Table 18 Soap Plants

Natural soap and washing plants. Plants that have a high saponin content make a lather with water. Other plants are healthy washes for various purposes.

Horse chestnut	*Aesculus hippocastanum*	P Soap plant
Agaves (US)	*Agave americana*	TP Soap plant
Sweet chestnut	*Castanea sativa*	P Hair washing
New Jersey Tea	*Ceanothus americanus*	P Bathe for skin complaints
Lawn chamomile	*Chamamelum nobile*	P Hair tonic
Fat hen	*Chenopodium album*	A Soap plant
Soap lily	*Chlorogalum pomeridianum*	P Soap plant
Ivy	*Hedera helix*	P Soap plant/hair washing
Honeysuckle	*Lonicera ciliosa*	P Hair washing
Ragged robin	*Lychnis flos-cuculi*	P Soap plant
Dwarf mallow	*Malva pusilla*	A Tooth cleaner
Wild chamomile	*Matricaria recutita*	A Hair tonic
Pokeweed (US)	*Phytolacca americana*	P Soap plant
Poplar	*Populus trichocarpa*	P Soap plant
Soap-bark tree	*Quillaja saponaria*	TP Soap plant
Rosemary	*Rosmarinus officinalis*	P Hair wash
Soapwort	*Saponaria officinalis*	P Soap plant
Soapberry	*Shepherdaria canadensis*	P Soap plant
White campion	*Silene alba*	P Soap plant
Red campion	*Silene dioica*	P Soap plant
Soapweed	*Yucca glauca*	P Soap plant

from the house can be split into two levels of quality. Grey water is water that has been used for washing and therefore is usually derived either from the bathroom or the kitchen sink or washing machine.

Apart from tap water ingredients, this will contain dirt from clothes and human washing and the soaps and detergents used in those processes. Clearly if ecologically safe soaps and washing powders are used then the quality of the water will be less harmful. It is important not to use untreated water in the garden when it contains substances such as household bleach. Soapy water can be used directly in the garden for irrigation, provided it is not put on to plants which are likely to be ingested directly, such as small fruit and salad plants.

In Chapter 3 we looked at systems for treating waste water. This is certainly a necessary process for sewage water, if that is to be reused in the garden. There are also relatively straightforward means for recycling grey water through cleansing systems.

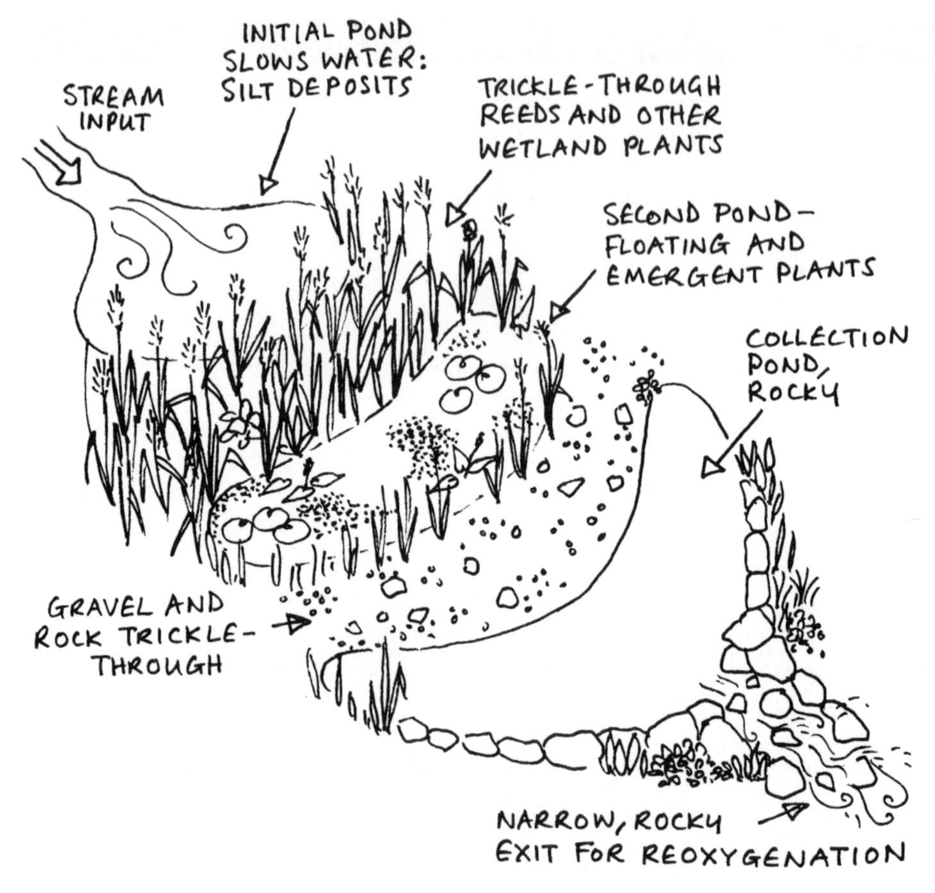

'Biological' washing powder is actually full of chemicals. Here is a true biological water cleansing system for fresh water streams.

In dry zones precipitation can be increased by making greater areas for condensation, and the interception of airborne moisture such as is available. This is a natural function of trees. However, structures such as corrugated iron roofs are particularly useful in having a good capacity to form condensation.

The gardener who has fresh running water from streams or rivers adjacent to their plot is lucky indeed. There are usually regulations limiting the interruption of natural water courses and anybody proposing to do so should be aware of how local legislation affects their rights in this regard.

You also need to be clear about the quality of water reaching your land from elsewhere in the watershed. Streams which are choked with green growth have become 'eutrophied'. This

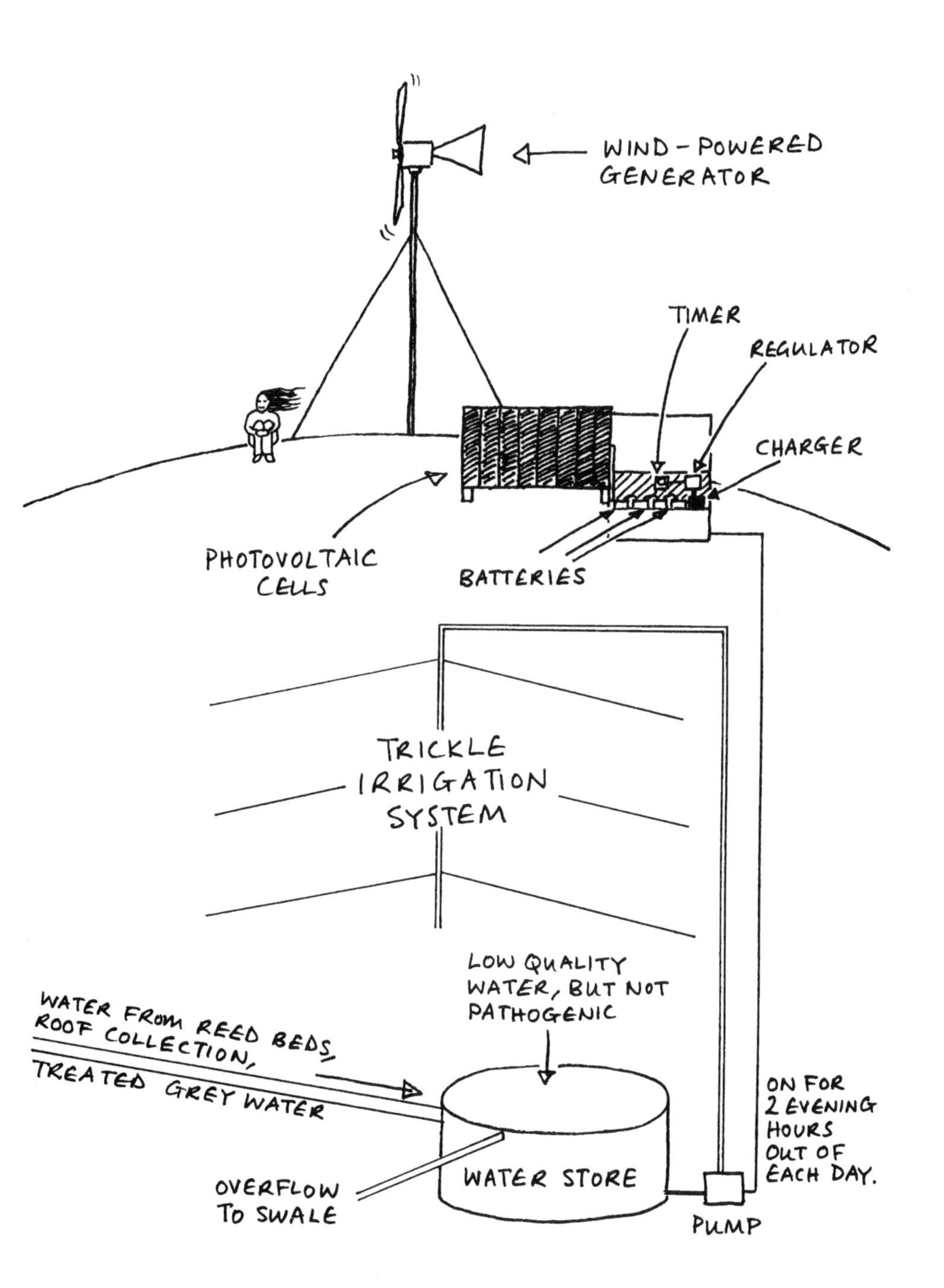

Water distribution can be either active...

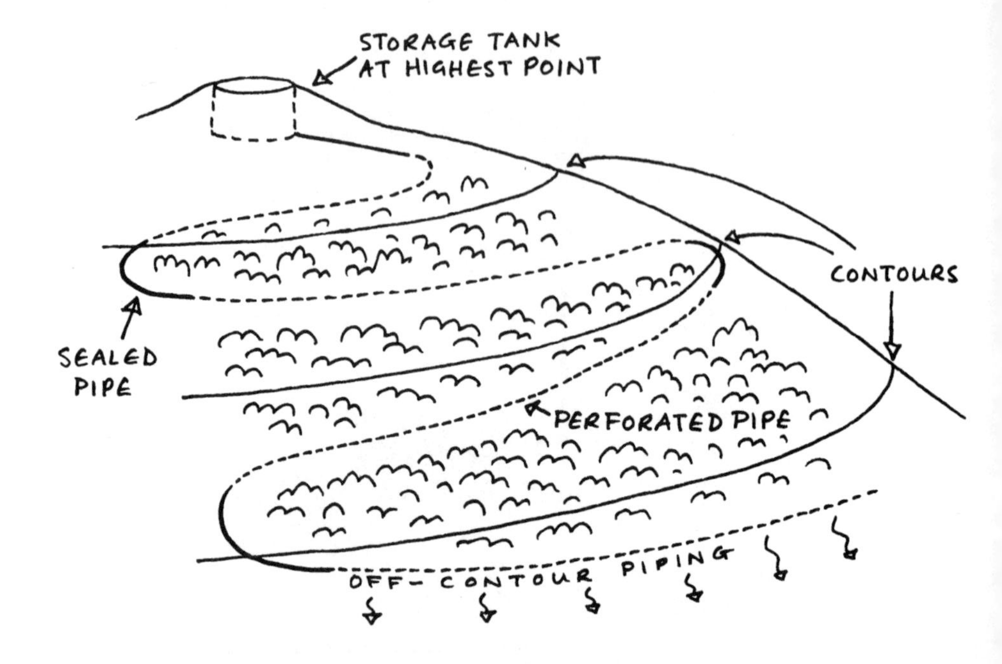

...or passive.

means they have excess nitrogen, caused by agricultural run-off, usually from chemical fertilizers. The water will have little oxygen in it. You can harvest the plant matter as green manure, either for mulching, or the compost heap. You can also clean up incoming water supply by the same biological processes used to clean outflow.

As we have seen, it is important to have adequate storage for water, and the volume of water available for storage can be calculated quite easily. In the case of drainage from roofs, this will equal the horizontal surface area of the roof multiplied by the annual rainfall (the roof pitch does not alter the area of water collection, which is the same as the ground plan of the roof).

We saw earlier useful devices to intercept downpipes from roofing gutters so that excess water can continue to drain once the water butts are full. It is also possible to construct tanks from reinforced concrete and a whole range of scrap material. Remember, water should be stored at the highest point possible in the garden so that the force of gravity can be used to work any irrigation systems.

It is also possible to use wind-powered pumps, and there are a range of low-energy water pumps available, which generally consume little more electricity than a light bulb. However, if by suitable placement we are able to run water around the garden simply by gravity then even this small cost can be avoided.

Given the state of most urban water supplies, no sustainable gardener should be looking to rely on mains water for garden irrigation in future.

The building of healthy top soil with high humic content is also one of the best things that the gardener can do to store water where it is immediately available to plants in the garden. Mulching and continuous ground cover are good techniques for reducing water loss by run off and evaporation. In turn they tend to add to the humic content of the soil, increasing its long term capacity to hold water.

Lastly of course the biomass in the garden itself is very rich in water, so if more plants and trees inhabit the garden area, the greater the storage of water within its growing structure will be.

Using it as a Medium for Yield

Water offers a wide range of yields directly and indirectly. It has species that are adapted to different micro-climates just as on the land. There are, for instance, those plants and fishes that prefer to live on the bottom of a pond, those that like to be in shallow water, and those that do best at the edge of a stream or pond.

Additionally there are a few plants that do not attach to the soil at all but simply float on the surface of the water. Amongst these are to be found the top feeding fishes. As with the forest or natural grass land, a healthy water base system is a natural balanced ecology. Bottom feeding fish (such as carp) clean up detritus that sinks to the floor of the pond.

Surface floating plants (e.g. duck weed and water lilies) help shade the pond and keep it at an even temperature. This helps maintain oxygen quality in the water. Oxygenation of water occurs at surfaces where fish and insects are feeding, through the transpiration of plants and also where water is flowing downwards over surfaces: the addition of waterfalls to circulatory systems (any moving water) in garden water is therefore vital in restoring the oxygenated quality of the water.

Water has other indirect yield, such as its ability to reflect sunlight, either to lighten a dark corner in the garden, or to reflect low winter sunlight onto the sun side of a house. Water takes much longer to warm up and then to cool down than earth, and therefore large water storages act as stabilizing influences against the vagaries of climate. Large underground tanks of water can also be used to store solar energy from the summer season through into the winter to be released to heat the house (see *Living with Energy*, in Booklist).

With an adequate supply of water there is also the possibility of generating power from water wheels. There is a whole range of possibilities for using this, from low-energy options such as turning a potter's wheel through to the generation of electricity . The power of running water can also be used to pump part of the water back up to levels higher in the garden.

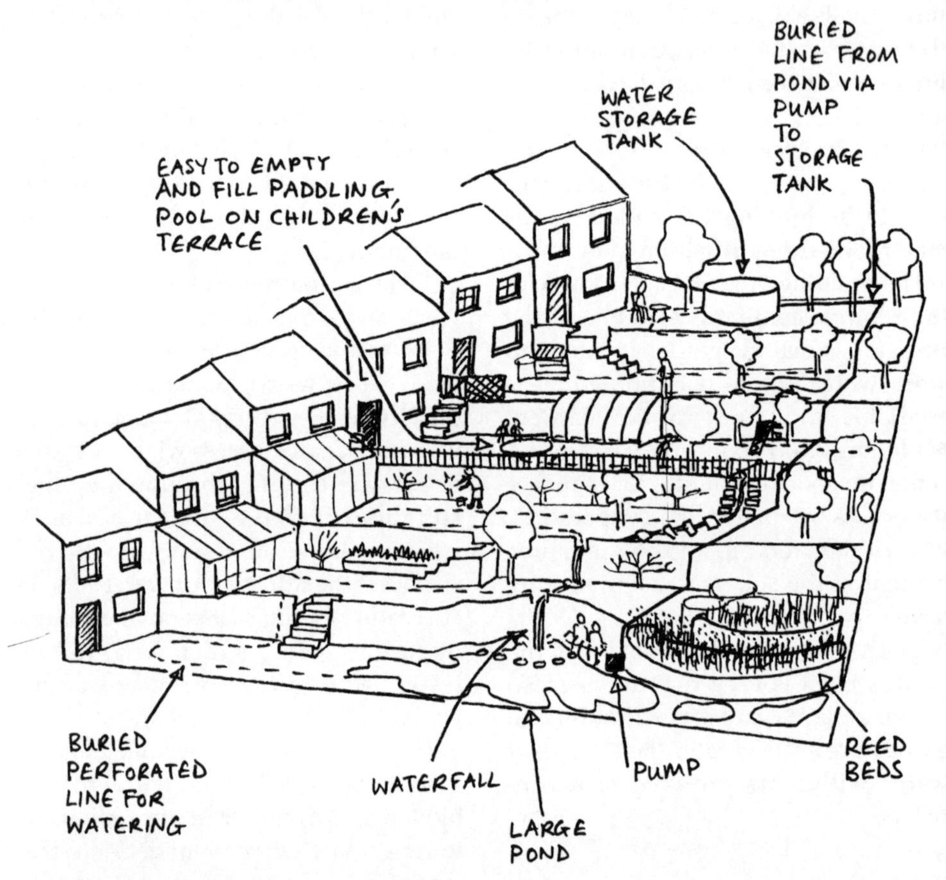

In a big garden, or where neighbours will cooperate, water can be used as a proper circulatory system, bringing a feeling of life to the garden.

Reed Beds and Grey Water Systems ~

We saw a design for a reed bed in Chapter 3. Reed beds can be used to create biological systems for cleansing water supplies. Reeds and other wetland plants have root structures which are adapted to anaerobic conditions, that is, where there is very little oxygen in the ground owing to the waterlogged nature of the soil.

Plants which are able to function in these conditions are very good at trapping and detoxifying heavy metals, chemical pollutants such as petrol derivatives, and any toxic waste in materials such as human sewage. They do this by microbial associations at root level which trap these undesirable compounds by turning them into bound chemical forms.

An area of reed bed with as little as one square metre per person is adequate to deal with that person's total waste water output. Beds are constructed of layers of rock, gravel and sand from coarse through to fine, and are planted up with native species of wetland plants. The diagram shows one such system.

By designing a series of beds it is possible to offer primary, secondary and tertiary treatment of water in different stages. Such systems have been described as 'engineered marsh land' and they hinge on the symbiotic design of plant, soil and moisture in a beneficial relationship. Where higher volumes of through-put are required it is better to construct a series of small systems rather than one big system; this gives greater energy efficiency and reserves in the event of breakdown.

The primary stage of the system is designed to remove solid waste. The secondary stage enables the breakdown of ammonium into nitrate material, and the third phase removes these nitrates and any phosphates. The system becomes more aerobic as it progresses, and in the more aerobic conditions more efficient bacteria exist. Ideally the final water output should run through a reconditioning system to aerate the effluent water.

Some people maintain that such water should not be reintroduced to systems for human consumption, but should be allowed to drain away to natural circumstances. However it is worth noting that water cleansed through a reed bed system is of considerably better quality than that put through municipal sewage systems, which are often recycled into the drinking water supply. Reed beds can be designed as gardens in their own right and are entirely smell free, as well as looking extremely attractive.

As will be seen from the areas quoted for treatment per person, household supplies can be dealt with in quite a small garden space. Obviously there will be planning regulations about whether or not one can treat sewage this way in urban situations, and anybody intending to treat their sewage through a reed bed system should ensure that they have appropriate professional advice before embarking on such an experiment.

In practice it is thought that such systems may work best with groups of people larger than the family. The design of small sewage treatment systems servicing a number of houses may be the way forward. This is certainly the experience in China, where millions of backyard sewage digesters have been built. Here, they prefer 'methane digesters', which turn the waste into compost, trapping and yielding as fuel the methane gas.

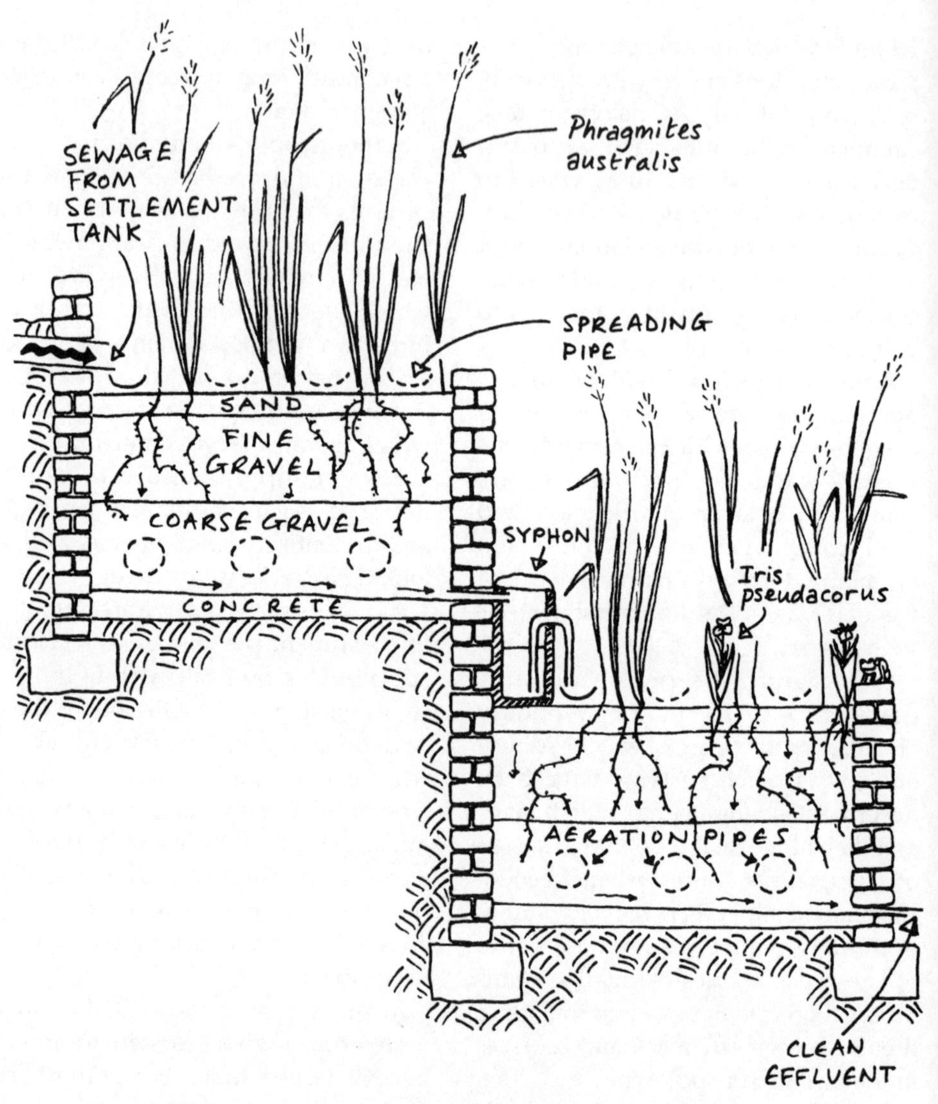

Water treatment emulating a natural marshland.

Reed beds have been sufficiently successful to operate on town scale, and a number of the municipal water authorities in Britain are now moving in the direction of designing such biological cleansing systems. Indeed they are already in use on sites of several acres by chemical companies to clean up industrial effluent.

It is worth remembering that the average anaerobic bacteria is as lazy as the next person, and will not wear itself out eating up highly toxic material if you give it something as nutritious and easy to get hold of as sewage. Therefore it is important that if dealing with toxic waste it is separated out from richer fare so that the plants will be able to trap it appropriately.

There are a number of consultancies growing fast around the world which are dealing in biological sewage and waste management systems, and we can safely say that this will soon be seen as a sustainable waste management system of the future.

Our present habit of thinking of sewage as something unhygienic which 'must be got rid off', is gradually being turned around to the realization that human effluents are full of nutrients and that policies that simply pump them out to sea not only pollute the environment into which the wastes are diverted, but furthermore are highly wasteful,

It is also possible to treat human sewage through dry management

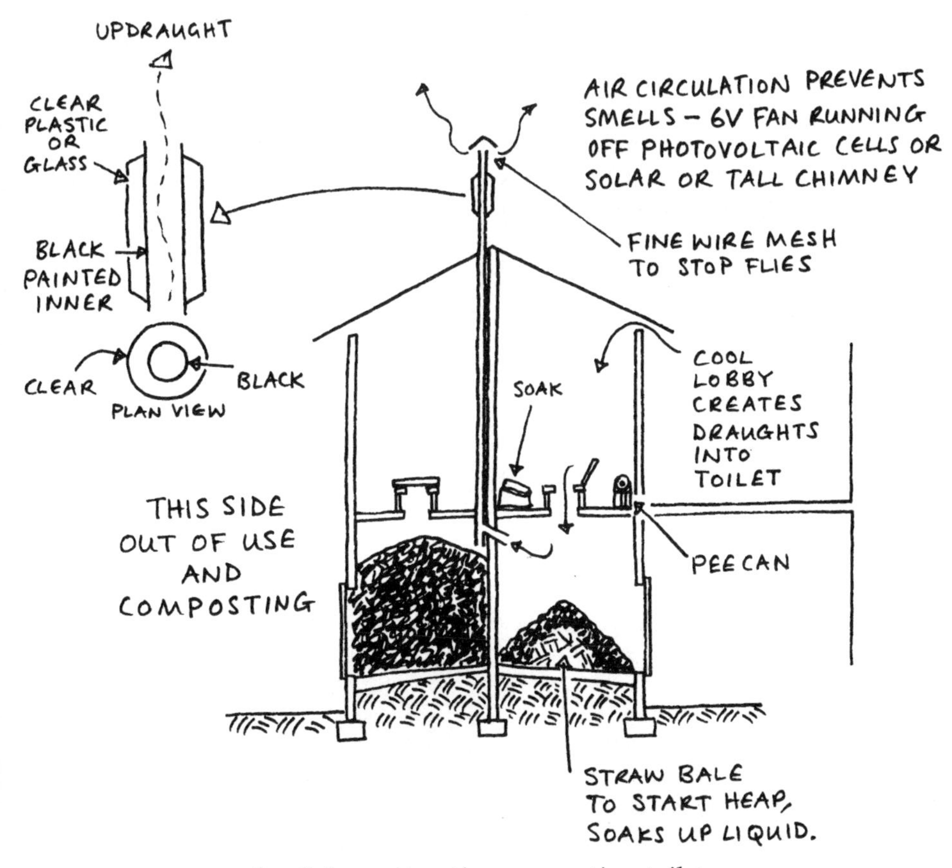

Smell-free, attractive composting toilet.
The complete fertility cycle for the garden.

systems. Here the waste is kept in enclosed composting systems and allowed to degrade over periods of time. Dry managed compost systems have the advantage that they take no water, although they must be carefully designed to ensure no seepage of effluent into areas where there would be a risk of cross infection.

The diagram shows a system built by Steve and Yvonne Page in France, which is permanently sustainable. Many other such systems have been built. It has twin closets: one is in use whilst the other one is sealed, digesting its contents. It is important to have an air-lock between the toilet area and the warm air of the house to avoid drawing smells up from the pit into the house. A proper design like this will produce rich fibrous compost in as little as six months. This can be applied straight onto the garden, is odourless and has all the appearance of any other organic fertilizer.

There is an immense satisfaction when you create such a system and realize that you have repaired a long-broken link of nature. You turn about from piped sewage, where the goodness of the land passes through the human gut to be swept (irretrievably) out to sea, to a renewable system of trapping and reusing fertility. Dinner guests are encouraged to visit the facilities often, as you realize that every offering is an addition to the potential yield of your plot. With careful design the toilet taboo can be broken and turned into risk-free abundance!

nine

The Forest Garden

Is it too much to hope...that the Iron Curtain of the world will give place to the Green Front and the scars of the earth as well as the scars in people's hearts may be healed by tree planting.

RICHARD ST BARBE BAKER

Imagine yourself in a wild woodland on a sunny summer's day. Parts are densely shaded, elsewhere sunlight plays on greenery in the clearings. The birds sing, while the undergrowth rustles with life. All around buds are bursting and the ground swells with vigorous young leafy plants. Flowers and fruit form at all levels around you. This is the original garden of Eden. It could be your garden too.

Putting It All Together

The logical conclusion of the sustainable garden is that it becomes a forest garden. Millions of backyards linked up across the continents to become forests filled with people, abundant in food, fibre and fuel. Urban wastes returned to greenness. Each individual gardener who makes a forest garden contributes significantly to the realization of this dream. There are a half a million hectares of back gardens in Britain alone. At only a hundred trees to the hectare that is a possible forest of fifty million trees!

All the techniques for soil creation, healthy relationships between plants and the use of multiple dimensions, bring us back to the natural woodland as our model.

The forest garden is not, however, a wilderness. It is a carefully managed space maintained for human benefit. The vertical dimension is used as much as possible to increase yield from the available area. Plants are selected to flourish through all the seasons. Perennials and self-seeding annuals are favoured. Trees are chosen for cross pollination and cropping through the seasons.

Table 19 Fruiting Plants

An enormous range of trees, shrubs and smaller plants can be cultivated for their fruit.

Serviceberry	*Amelanchier candensis*	P
Snowy mespil	*Amelanchier lanarkii*	P
Strawberry tree	*Arbutus unedo*	TP
Bearberry	*Arctostaphyllos uva-ursi*	P
Barberry	*Berberis vulgaris*	P
Ornamental quince	*Chaenomeles japonica*	P
Hawthorn	*Crataegus* spp	P
Azerole	*Crataegus azarolus*	P
Quince	*Cydonia oblonga*	P
Russian olive	*Eleagnus* spp	P
Fig	*Ficus carica*	P
Alpine strawberry	*Fragaria vesca*	P
Gaultheria	*Gaultheria shallon*	P
Huckleberry (US)	*Gaylussacia* spp	P
Sea buckthorn	*Hippophae rhamnoides*	P
Japanese raisin tree	*Hovenia dulcis*	P
Mahonia	*Mahonia* spp	P
Apples	*Malus* spp	P
Medlar	*Mespilus germanica*	P
Mulberry	*Morus nigra*	P
Cape gooseberry	*Physalis edulis*	A
Cherry	*Prunus* spp	P
Apricot	*Prunus armeniaca*	P
Plum	*Prunus domestica*	P
Peach	*Prunus persica*	P
Sloe	*Prunus spinosa*	P
Pear	*Pyrus communis*	P
Sumach	*Rhus* spp	P
Roses	*Rosa* spp	P
Gooseberry	*Rubus grossularia*	P
Black currant	*Rubus nigrum*	P
Red currant	*Rubus rubrum*	P
White currant	*Rubus sativum*	P
Arctic brambleberry	*Rubus arcticus*	P
Cloudberry	*Rubus chamaemorus*	P
Blackberry	*Rubus fruticosus*	P

Raspberry	*Rubus idaeus*	P	
Loganberry	*Rubus loganobaccus*	P	
Japanese wineberry	*Rubus phoenicolasius*	P	
Elderberry	*Sambucus nigra*	P	
Red berried elder	*Sambucus racemosa*	P	
Whitebeam	*Sorbus aria*	P	
Rowan	*Sorbus aucuparia*	P	esp. *var edulis*
Checkers	*Sorbus torminalis*	P	
Blueberry (US)	*Vaccinium corybosum*	P	
Cranberry (US)	*Vaccinium macrocarpon*	P	
Bilberry	*Vaccinium myrtillus*	P	
Cowberry	*Vaccinium vitis-idaea*	P	

Shade-loving plants occupy darker corners, while there is space left for those that require sun. Trees are used to make warm pockets for trapping maximum growing advantage in the summer, and sheltering more delicate perennials through the winter.

As the trees grow and the underlying fertility of the plot increases, the need for maintenance actually decreases. There will be more leisure time to enjoy the product of the garden and to perfect its output. Contrast this with the garden which is a pressure because 'there is always work to do'!

The Architecture of Trees

To use trees well in a garden it helps to understand something of how they work. There are also ways in which human adaptation has been used to increase the usefulness of trees. Trees are the ultimate perennials in the garden. Knowing the possibilities for their use gives a strong yet flexible framework over which the rest of your garden design can be secured.

When you look at a tree, what do you see? Trunk, branches, buds, leaves, flowers, fruit....Yet half the tree is hidden from your sight. In planting and caring for trees it is important to remember this invisible side of the plant and to protect and feed it well.

In our contemporary hurried world there are so many sensory stimuli that it's easy to forget the unseen. When people talk about 'getting back to their roots' they are feeling the need to pay attention to this unseen part of their own being. Without this connection no living organism can flourish.

Tree root systems are branched just like the upper part of the tree. Apart from firm anchorage this is crucial for feeding the whole. Branching is a natural phenomenon that occurs in many living structures. Think of the human circulatory system as an example, or of following a large river back to its many sources. Branched structures are well adapted for the flow

The tree, the whole tree, and nothing but the tree...

of liquids. Trees are superbly evolved in this way to pass food in solution to all the parts of their 'bodies'.

Branching gives the tree a huge surface area compared to the space it occupies at ground level. This means the tree has as much opportunity as possible to interact with the air and the soil. Trees have evolved to absorb energy above ground by converting carbon dioxide to sugars, by photosynthesis in their leaves. Below ground their root hairs enable them to absorb nitrogen, water and other essential minerals from the soil.

Forests, then, are the lifeblood of the planet in two ways. They take carbon dioxide from the air, trap it in their biomass and release oxygen ('photosynthesis'). They make it possible for human life to take place, because we need to breathe oxygen and exhale carbon dioxide. They are in effect reversing a cause of the greenhouse effect. More trees means less problem. Only the vast unfathomable oceans have such an effect on keeping the balance of our air endurable for human life.

Secondly, the deep exploratory roots of trees break up rocks below the soil level, and release minerals needed to make topsoil good enough for other plant life. Fine root hairs absorb these minerals in solution. The tree is a living pump, moving this food around its structure through vessels near the surface. This is the sap which pours out when you wound a tree. If you have ever cut down a large tree then you'll have seen the sap being pumped from the stump for some time after the felling. This shows you the fantastic flow rate trees achieve.

When the tree loses its leaves, they fall to the ground and make the goodness available to other soil and plant life. In turn the roots of the tree, and the windbreak it provides, tend to stabilize soils and prevent erosion and run-off. Trees are thus vital to the building of topsoil, and in turn the whole planetary cycle of fertility.

When you see the 'rings' in a piece of wood cut across the trunk, you are looking at the annual cycles of growth over past years. Trees grow quickly in the warm seasons, and more slowly in the winter, giving a marked grain. As each year comes along the wood gets thicker. Only this year's growth is at the outside of the trunk. So in effect the greater part of the tree is actually dead material. It is only the *cambium* or living outer layers that keep growing.

This core of hard woody material is what gives the tree strength, and in time gives us all our useful timber products from furniture to firewood. When the tree dies and is left to rot it all decays back into the soil to help build deep black, humicly rich 'forest soils' which form some of the best growing land we have.

Over thousands of years we have developed selective propagation of trees, choosing those which yield best, and so developing 'varieties' or 'cultivars' with especial virtues. If trees are grown on from seed then they are subject to genetic change. You cannot predict the exact characteristics of such a tree until it grows up. By then many years will have passed, and if it is not especially fruitful then time will have been wasted.

Pruning and Grafting

All this hit-and-miss process can be avoided by using *vegetative propagation*. Some plants will 'take' by pushing cuttings into the soil. When the cutting roots, a new plant grows which is genetically a copy of the donor plant. For trees a more involved process is usually used.

Living material is taken from a tree which has the desired characteristics and is grafted onto an existing rooted tree. The end result is predictable. In this human-designed tree the rooted part is called the 'stock', and the top part is the 'scion'. Grafting is fairly straightforward, and anyone can consult a simple manual and have a go. Success comes with practice.

In reality most people will buy their garden trees ready grafted. It is still useful to understand the process, however, to know what you are buying. The type of stock used will govern the vigour of the tree, and its ultimate height. Small gardens should use dwarf or semi-dwarfing stocks, larger gardens might go as far as half-standards. Only full-scale orchards will have room for standard trees.

In practice most commercial orchards now prefer smaller stocks, as it makes picking very easy, you can get more

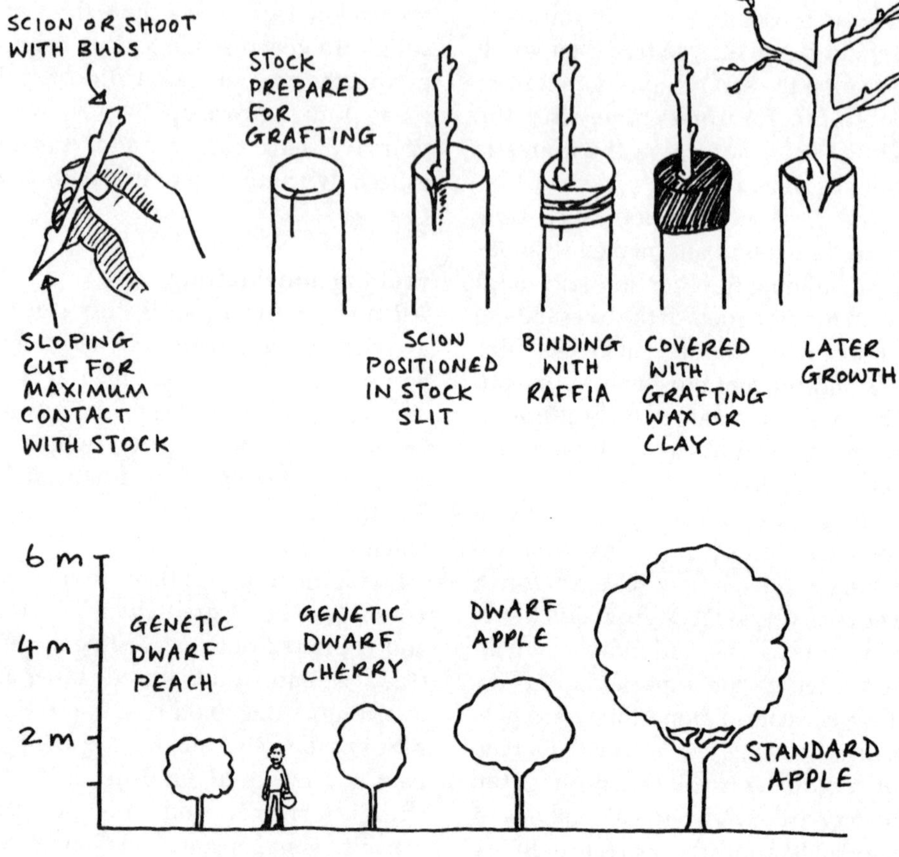

Vegetative propagation of trees.

trees to the hectare and the trees come into fruit more quickly. Cultivars on bigger stocks are slower to fruit, require less maintenance over their lifetime and live longer.

We can also alter the shape of trees and increase their fruitfulness by pruning.

These shapes of tree can all be used to grow up walls, or to limit the tree to two dimensions, making living trellises between garden areas. The neglected art of step-over espalier is particularly useful for getting fruit from the edge of vegetable plots. Not all trees will accept all of these shapes; apples, for example, do well as cordons and espaliers, whilst cherries prefer to be fanned.

Lastly, when discussing the architecture of trees we need to acknowledge that no tree can be seen in isolation from its surrounding environment. If the bee pollinates the tree, or the worm helps feed it, how can we say that the bee and the worm are not in themselves part of the living process of the tree. It is for this

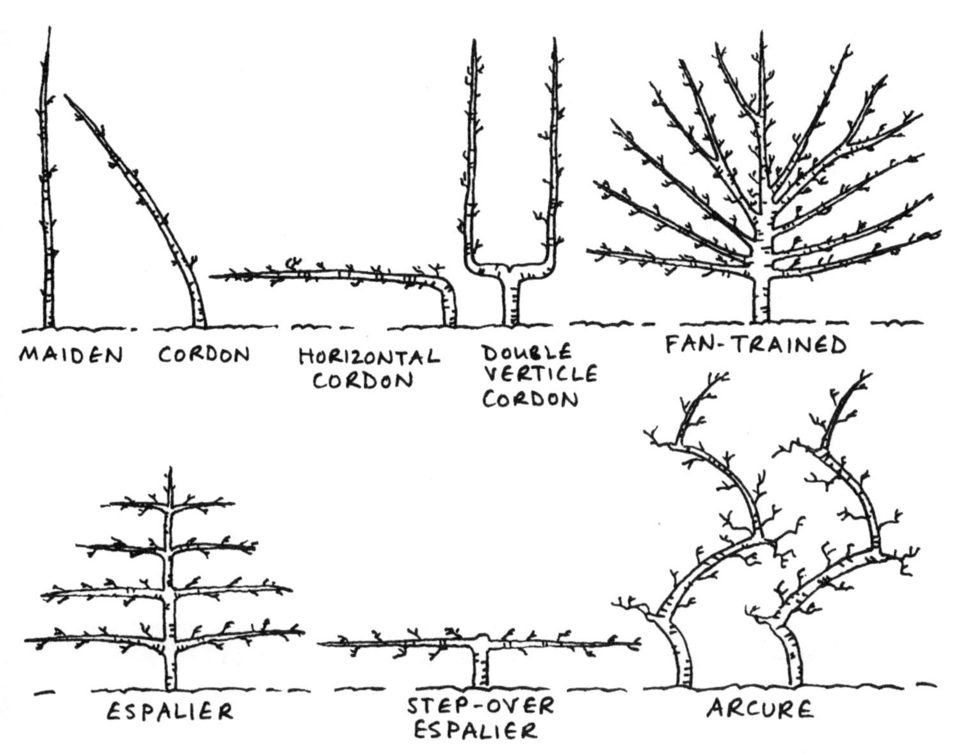

Trees pruned to shape give high productivity from small land areas.

reason that a well-designed sustainable garden will inevitably move towards the integrated structure of a mini-forest.

Tree Crops ~

Trees have many possible yields. Here are some:

- sap
- leaves
- flower products
- fruit
- thinning wood
- timber
- other plants

Let's look at the edibles first. Sap can be taken from certain trees to make wine or sugars (e.g. birch or maple). Leaves can in some cases be eaten, or made into wine (oak leaf wine, for instance). Flowers can be picked (e.g. lime) for tea. They also provide food for the honey bee. Fruit might be 'top fruit' (e.g. apples, pears, plums) or nuts and seeds (e.g. hazel or walnut) destined for human consumption; alternatively they might be collected or free range grazed for animal feed (e.g. acorns, beech mast, mulberries). Human and livestock systems can be integrated, so after the apple harvest pigs can be used to clean up windfalls and give the orchard a quick manuring (although, as I mentioned earlier, pig manure imported from intensive farms is *not* good for the garden as it contains chemical growth promoters etc.).

Moving on to wood harvesting, thinnings can be gathered, whether twigs from the woodland floor, loppings from large trees, or thin wood from whole trees taken out of woodland to make more room for selected trees to grow on. Timber is larger wood gathered when mature trees are felled. The more we can make do with thinnings for our wood needs, the less we will need to cut down big trees.

Trees which regenerate after partial cutting are particularly useful. Hazel, sycamore, ash, and hornbeam coppice very well; that is, they can be cut off at ground level and the old stump will usually regenerate by throwing out new branches. Cycles of coppice can be set up over five to twenty five years, depending on site and tree. Willows are coppiced or pollarded. Pollarding means cutting off branches at head height, a technique used well on limes (*Tilia* spp) and plane trees.

Many other plants will thrive in woodland. In Chapter 2 we looked at maximizing the use of all layers of growing space from roots to high trees.

Table 20 Unusual Roots

Some less common root crops

Common name	Latin name	Use
Garlic mustard	*Alliaria petiolata*	A Horseradish substitute
Greater burdock	*Arctium lappa*	B Boil as vegetable
Horseradish	*Armoracia rusticana*	P Grated as relish
Pleurisy root	*Asclepias tuberosa*	P Lung tonic
Red valerian	*Centranthus ruber*	P For soups
Chicory	*Chicorium intybus*	A Coffee substitute
Earth almond	*Cyperus esculentus*	P Dessert/coffee
Yellow gentian	*Gentiana lutea*	P Bitter tonic
Jerusalem artichoke	*Helianthus tuberosus*	P Boil as vegetable
Bog bean	*Menyanthes trifoliata*	P Edible root
Oca	*Oxalis tuberosa*	P Boil as vegetable
Evening primrose	*Oenothera erythrosepala*	B Use first year roots as herb
Common reed	*Phragmites communis*	P Dry and grind rhizome for flour
Chinese artichoke	*Stachys sieboldii*	P Boil as vegetable
Dandelion	*Taraxum officinale*	P Coffee substitute
Goat's beard	*Tragopogon pratensis*	B Parsnip substitute
Tuberous nasturtium	*Trapaeolum tuberosum*	TP Boil as vegetable
Reed mace	*Typha latifolia*	P Boiled/grated vegetable

Tree crops will include the yields of climbers (e.g. blackberries, ivy pollen) and epiphytes and parasites (mosses, ferns and fungi).

When economy was firmly rooted in the produce of woodlands, people developed a massive fund of knowledge on the properties of all these plants. Their health-giving benefits, their value as weaving and dye plants, and their use as fuel stock were all fully appreciated. In consequence their supply was much sought after, and in feudal societies severely governed by statute. Our surprise at discovering the amazing resourcefulness of Amazonian Indians in their tree-based culture should really be an astonishment, not at their wisdom but at our own ignorance. We had once and have now largely lost this knowledge which was so recently integral to our own societies.

Whatever will be useful in the high woodland can be understood and utilized also in your own back garden. As has been suggested before, start small and then increase the understanding and use you make of this aspect of the sustainable garden. Learning one new species of plant a week will give you fifty in a year. It doesn't require huge effort to soon build prodigious knowledge. Once you recognize a particular plant you can start to learn its uses. I particularly recommend herbals for this (see Booklist).

If the same amount of energy had gone into developing tree crops as has been given to vegetable or grain varieties in the last hundred years, what a wonderfully productive forest system we might have! As we learn to appreciate trees afresh, then let us hope that research will move in this direction.

Trees as Interceptors

Trees are wonderful interceptors of all sorts of free energy in the environment, trapping it and making available resources for human consumption. That trees interact with the soil and underlying bedrock we have already seen. In addition, trees:

- turn sunlight into stored fuel
- attract insects and wildlife
- make windbreaks, and convert wind action into 'soil stirring' through their roots, aerating the earth for other plants
- collect downhill run-off of organic matter
- increase and trap precipitation

As an example, consider rainfall. By constant transpiration trees draw up water from the water table below ground and put water vapour back into the air. A mature deciduous tree may evaporate five hundred litres of water on a warm summer's day. All of this water is then available for further precipitation, adding to the cycle of life.

In turn, when it rains, the tree will intercept and collect rainfall. Some seeps into the tree, some evaporates without ever reaching the ground. That which does reach ground level tends to concentrate around the trunk or at the 'drip line'. This is a combination of the umbrella effect of the leaves shedding the water outwards, and the internal branches draining back down the central stem of the tree.

In planning the garden be conscious then that these two areas are wetter, and that below the main spread of branches is comparatively dry.

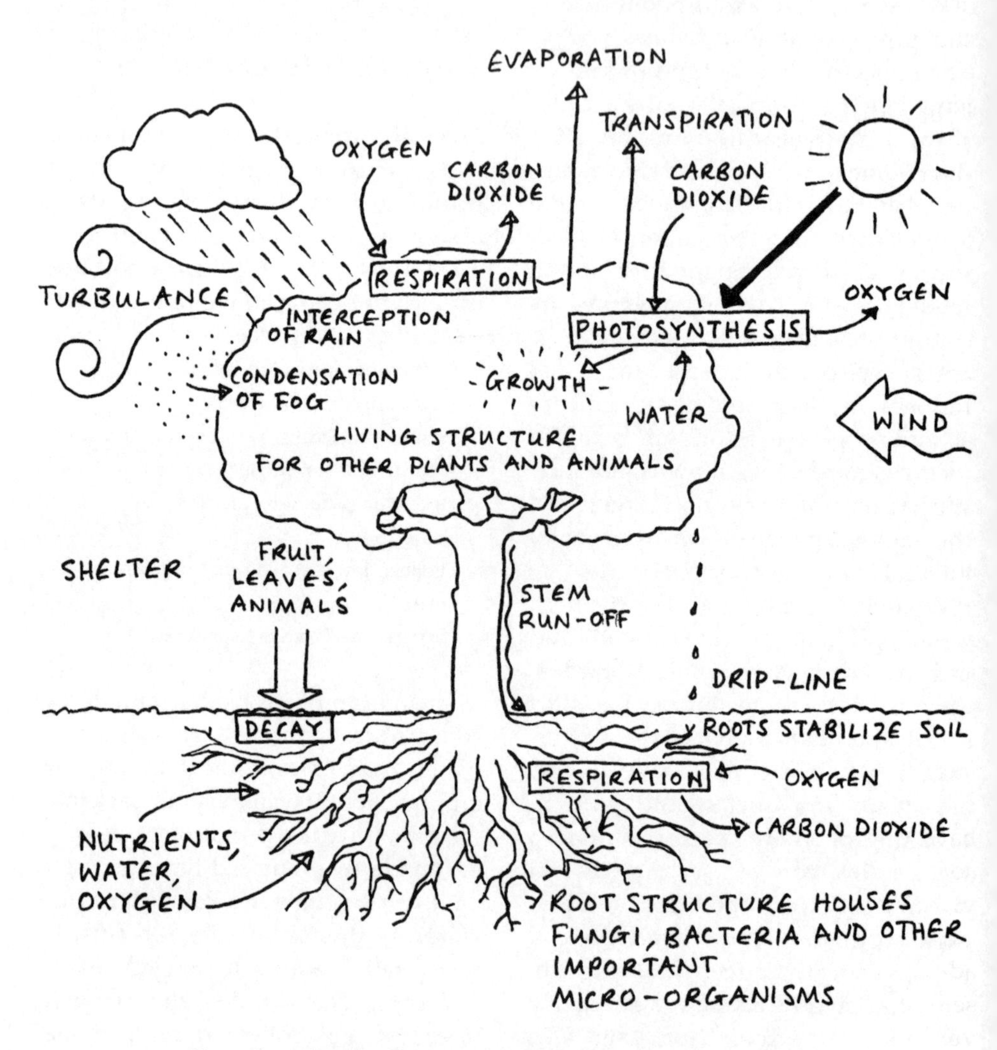

Trees benefit us all by their ability to trap energy from the environment and turn it into useful products.

Other Yields

Perhaps the greatest yield of trees is what they do for the human soul. To be in delightful woodland is one of the most enchanting experiences on the planet. Industrial society is very people-centred; in the forest you get a sense of life and scale beyond human interests and needs which reminds us that we are part of a living world much greater than ourselves.

Trees in a garden bring a little of this feeling right up to our back door. They give us shade and peace. They invite us to rest. In the wind and rain they roar and make evident the power of natural forces.

Trees also equalize extremes of climate. Their presence on the forest scale cools hot climates and warms colder ones. This you can also experience in your garden. Suitable wind-hardy species will cut out the penetrating winds which often whip round corners of houses. They also provide shade in sun-trap gardens to cool a languid summer's afternoon.

Woodland and permanent pasture have high levels of earthworm activity, so more soil is being created in these places.

Blossom, leaf colour and bark tints all add to the beauty of the garden. Careful selection of trees for these features is very effective at brightening winter, spring and autumn in the garden. Birches and *Prunus* species can have particularly attractive winter bark. Beeches and limes are vibrant in spring leaf, liquidamber and maples very effective in the autumn. Good colour catalogues from garden centres and nurseries are full of this sort of information.

Relationships Between Trees

Trees do not exist in isolation any more than anything else in nature. It is interesting to observe native woodlands, to see how the wild wood builds itself as a community.

Some trees need to be at the edge of clearings, some can tolerate deep shade. Some (beech in particular) cast such a deep shade that little other than fungi can live under them.

Alder and willow, with bog myrtle on acid upland and alder buckthorn in lower climes are examples of trees whose role in nature is to reclaim wetland. By having respiratory systems which can survive in the anaerobic conditions of wet bog, willows and alders put down roots and stabilize the ground. The wetland succeeds to carr (damp former bog with incipient tree growth), and in time dries out to scrubby woodland. Many other trees are adapted to successions like this.

The evolution of trees as pioneers can be used for benefit in the garden, in this instance to dry out a wet corner. Beware that trees with wetland root systems (especially willows) can break up building foundations, so keep them away from the house! Another example would be to bring in nitrogen fixers to feed the soil. Think, for instance of interplanting alder, *Robinia*, or *Caragana arborescens* between fruit and nut trees.

Very vigorous trees can be used as windbreaks to enable softer species to get established in their lee. In time they can be felled or cut back for timber or mulch material.

Fruit Trees

The most useful trees to have in a garden are fruit and nut trees. If you choose the right tree then they can give crops for many years with only occasional maintenance. Every fruiting tree will need other trees to cross pollinate it. Some cultivars are described as 'self-fertile', which means they crop without a cross pollinator, but received wisdom is that even these are more prolific given a pollinating partner.

Not all fruit varieties flower at the same time, so it is important that trees chosen to work together actually flower at the same time. In the case of 'triploids' (which amongst apples includes such popular cultivars as Cox's Orange Pippin and Bramley's Seedling) the tree will need two cross-pollinators. Be careful in selecting fruit trees that they have adequate frost hardiness for your site, and that they can flourish at your latitude. Don't trust garden centres to know! Many such establishments in Southern Scotland sell the aforementioned varieties although neither of them are north-hardy.

Fruit trees can also suffer from biennialism. This is a tendency to flower and fruit every other year. This can be kept at bay by preventing the tree overcropping. If a tree is very successful at pollination time it's as well to remove excess fruit by late May or early June. This conserves the energy of the tree and helps prevent the exhaustion which brings on biennialism.

There is not space here to give detailed guidance in pruning, but as general advice note that autumn and late winter pruning encourages woody growth, whilst summer pruning encourages the formation of fruit buds.

Comprehensive tables of suitable fruit varieties are found in *The Fruit Garden Displayed* and *The Good Fruit Guide* (see Booklist). A fuller explanation of choosing fruit trees is given in my previous book, *The Permaculture Way*.

Robert Hart is the great inventor of the term 'forest garden' in this sense, and his pioneering work is to be enjoyed and emulated from his books on the subject.

ten

Community Gardening

There's not a pair of legs so thin, there's not a head so thick,
There's not a hand so weak and white, nor yet a heart so sick,
But it can find some needful job, that's crying to be done,
For the Glory of the Garden glorifieth everyone.

RUDYARD KIPLING (1911) *THE GLORY OF THE GARDEN*

Everything functions in communities. Our habit of dividing nature up into subjects for the purposes of description becomes unhelpful when we want to put our knowledge back together as a working whole. Regarding the garden as a thriving community is a good way forward to achieving sympathetic husbandry.

Guilds

The one day tree gardens (Chapter 3) are good examples of guilds. A guild is just a name to describe a community of plants which get on well together. In nature plant communities evolve which are specially adapted to the soil type and climate of their situation.

Different plants occupy the different niches of time and space to give complete soil cover through the seasons, and a succession of species towards the 'climax' ecology. This is usually woodland, although in some continental areas it may be prairie.

In the garden we interfere with nature. To put it another way, we cheat. We are not building natural plant communities but selecting plants to meet our own needs. Often these plants are highly selected to give better yields. The garden cabbage of today is several centuries of careful management away from its wild ancestor.

However, we can still respect the wisdom of natural selection and look to build guilds of plants in our garden which are cohesive and mutually beneficial. As an example, in the Andes (original home of the potato) native growers would not dream of cultivating potatoes alone. They might be interplanted with tuberous nasturtiums, or other plants; the combination gives greater soil and plant health.

To devise your own guilds (and anyone can) think of a particular plant you wish to grow. What time and space does it occupy in the garden? What other plants would go well with that one to give a year-round crop? What other plants would occupy the space left around the chosen plant?

If you want to plant apple trees, think of shrubs and ground cover plants to interplant. If it's root crops you're after, what herbaceous plants would go between? What would be a good succession plant to the present crop? Could other species be interplanted beneficially? Don't be afraid to experiment. Remember, no-one ever learnt anything by getting it right. Only mistakes teach us new ways forward.

Including Animals ~

A natural ecological community is a balance of all forms of life. That includes animals. Unfortunately, in a world fraught with different views of the morality of animal husbandry it's impossible to give advice without upsetting somebody! My view of ethics on this matter is that we all have our own, and the most ethical approach is to respect that difference.

Whatever the current politics of the issue, the biology of it is that natural systems include animals. To manage animals well we should ask ourselves what are their needs, characteristics and outputs?

Rabbits need space, food, water, toilet space and other rabbits. They like to graze, dig, and be amorous from time to time. They output pellets, more rabbits, pelts, meat and bones (but these last only if you terminate their rabbitness). Rabbits are actually a very compact and handy form of backyard stock rearing for the pot. I even knew a lady once who did it for the fur coats, and made them into hats. If, like me, that particular use doesn't appeal to you, fret not. You can still breed rabbits, chickens, ducks, geese, turkeys, goats, pigs, guinea pigs, dormice, or even sheep in the back garden for food.

Rabbits and chickens can be run in 'tractors' (small movable meshed frames) to clear the garden for you. Better still they can be run in chicken fodder gardens, where they can free range for their supper. This is much more in line with the idea of garden as community. Hens that just take up a corner of the garden which they scratch bare are as much a mono-crop as your local agribiz farmer.

Pigs and chickens are very effective at eating up household scraps and turning them into more dinners. The old village way was to share a pig with neighbours at slaughter time, and get a share of theirs in turn. This sort of barter is a strong builder of local community between people.

Community in the garden is developed by the integration of animals into the plant and landscape design as much as possible. Usually this means limiting the freedom of the livestock to some extent. Whilst the occasional slug forage by ten hens is fine, free range poultry and seedlings are mutually exclusive.

Free range pigs and any garden worth mentioning at all are out of the question. Goats and trees are pretty incompatible.

If keeping livestock, try to design contained gardens for them which grow appropriate feedstuff for self foraging. The stock may have to be rotated through a small number of forage areas, or kept off them altogether during establishment. At the end of the day you can produce a regime which is less work for yourself and healthier for the animals.

The most common animals in the garden in Britain are regrettably cats and dogs. I don't regard this as highly sustainable. Whilst cats and Jack

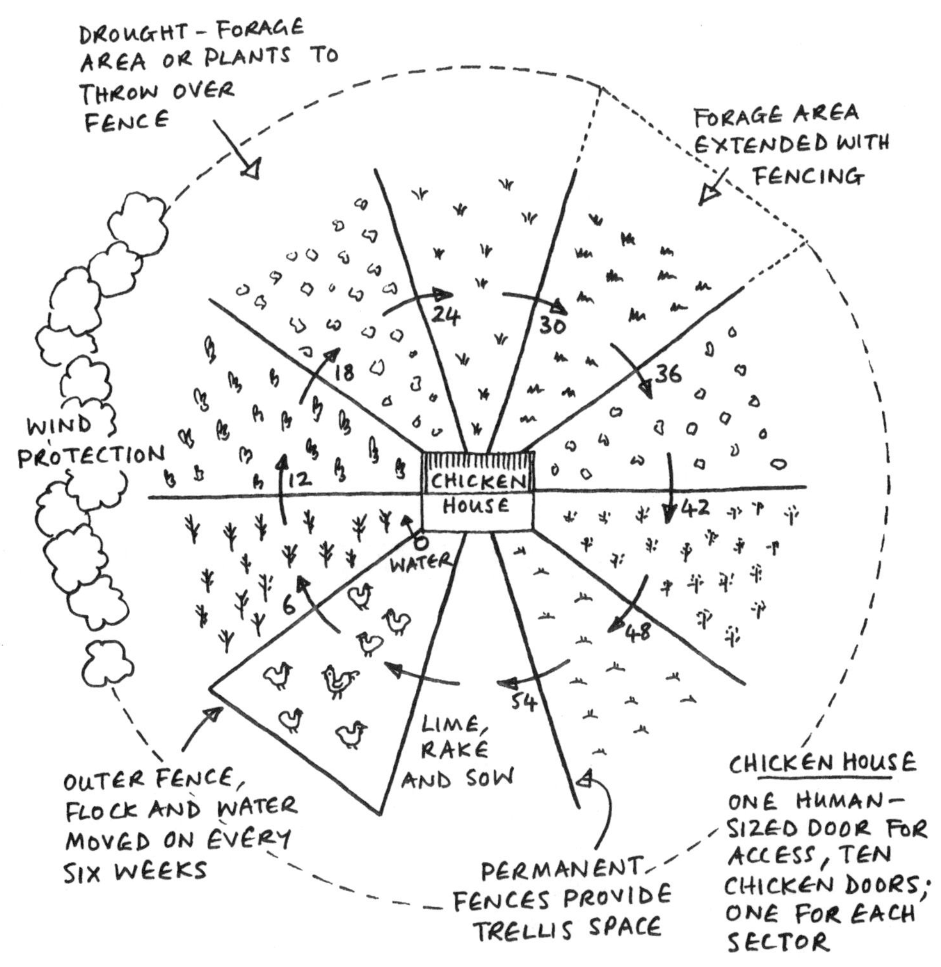

A design for a bigger garden with self feeding and watering, easily managed for year-round care.

Table 21 Chicken Fodder

Fodder plants for free-ranging hens. Chickens will eat just about anything. Hard seeds and green leaves which can be self-foraged are best.

Bamboo	*Arundaria macrosperma*	P
Shepherds purse	*Capsella bursa-pastoris*	A
Siberian pea tree	*Caragana arborescens*	P
Sedge	*Carex* spp	P
Fat hen	*Chenopodium album*	A
Good King Henry	*Chenopodium bonus-henricus*	P
Hawthorn	*Crataegus* spp	P
Broom	*Cytisus* spp	P
Beech	*Fagus* spp	P
Cleavers	*Galium aparine*	A
Jeruslaem artichoke	*Helianthus tuberosus*	P
Lupin	*Lupulinus* spp	A/P
Medics	*Medicago* spp	A/P
Wood millet	*Milium effusum*	P
Mulberry	*Morus nigra*	P
Plantain	*Plantago major*	P
Oaks	*Quercus* spp	P
Elderberry	*Sambucus* spp	P
New Zealand spinach	*Tetragonia tetragonioides*	P
Gorse	*Ulex europaeus*	P

Table 22 Animal Fodder

Fodder plants for grazing/browsing livestock

Horse chestnut	*Aesculus hippocastanum*	P	
Alder	*Alnus* spp	P	High protein
Sweet vernal grass	*Anthoxanthum odoratum*	P	Medicinal/fodder
Bamboo	*Arundaria racemosa*	P	
Siberian pea tree	*Caragana arborescens*	P	High protein
Sedge	*Carex* spp	P	
Sweet chestnut	*Castanea sativa*	P	
Chicory	*Chicorium intybus*	P	
Pampas grass	*Cortaderia selloana*	P	
Hawthorn	*Crataegus* spp	P	Horses like this

Animal Fodder Continued		
Broom	*Cytisus* spp	P High protein
Beech	*Fagus* spp	P
Goats rue	*Galega officinalis*	P Milk booster
Walnuts	*Juglans nigra*	P
Lespedeza (US)	*Lespedeza sericea*	P
Lupin	*Lupulinus* spp	A/P
Medics	*Medicago* spp	A/P
Melilot	*Melilotus officinalis*	B Medicinal/forage
Mulberry	*Morus nigra*	P Popular with pigs
Sainfoin	*Onobrychis viciifolia*	P For chalky/light soils
Gean	*Prunus avium*	P
Sloe	*Prunus spinosa*	P
Oaks	*Quercus* spp	P
Elderberry	*Sambucus* spp	P
Comfrey	*Symphytum* spp	P High protein
Gorse	*Ulex europaeus*	P High protein, crush for feed

Russells can be justified on the grounds of keeping rats and mice under control, the amount of money and attention lavished on pets seems obscene in a world where half the people haven't got enough to eat.

But there you are, I said the morality of keeping animals was a personal issue. The dog and cat lovers can accept the challenge of designing a garden community which includes their furry friends, and I'll stand corrected.

Water Integration

We've had a good look at water as an integral part of the landscape. In the sustainable garden standing or moving water will mesh with all the other elements of design. The way to think of this is as a continual flow of energy.

By way of example, note the following features in the illustration:

Water

- collection of run-off
- stabilizes temperature
- contains plants and nutrients

Surface Plants

- keep water cool
- provide habitat for surface feeding insects and fish

Edge Plants

- stabilize banks
- provide habitat for frogs and insects
- provide grazing for ducks
- green manure for garden

Bottom Plants

- utilize fertility of sunken trash
- give habitat for bottom feeders

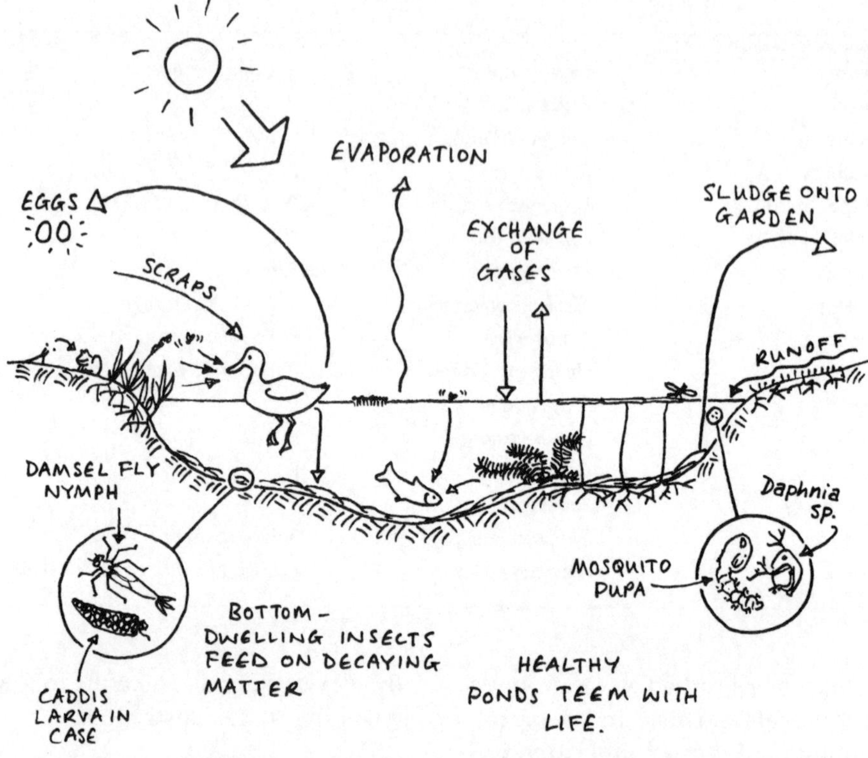

By respecting the natural energy flows the community of the garden becomes virtually self-maintaining.

Perch, brown trout and pike are predators, and will eat other fish. There are more edible freshwater creatures than these. Not all the fish in Table 23 are popular today. A sustainable garden pond will include the full range of species needed to set up lifecycles for the fish. That includes flies, larvae, beetles, worms and so on. The appropriate species for building a self-sufficient water ecosystem will depend on the locality and the stocking of the water.

Ducks

- keep greenery cropped
- fertilize pond soil
- provide outputs for human usage
- crop slugs etc. in the garden

Frogs

- keep down insects and slugs

Surface Feeders

- clear floating debris
- stop surface plants taking over

Bottom Feeders

- reuse fertility from surface feeders
- eat duck waste (turning it into fish)

Pond Sludge

- manure for garden when pond is cleaned

Table 23 Fish

Freshwater fish for the garden.

Common bream	*Abramis brama*	Bottom feeder
Eel	*Anguilla anguilla*	Bottom feeder
Swan mussel	*Anodonta cygnea*	Slow water
Crayfish	*Austropotamobius pallipes*	Chalky bottoms
Barbel	*Barbus barbus*	Bottom feeder
White bream	*Blicca bjoerkna*	Slow water
Crucian carp	*Carassius carassius*	Swampy lakes
Common carp	*Cyprinus carpio*	Bottom feeder
Pike	*Esox lucius*	Open water
Medicinal leech	*Hirudo medicinalis*	Ponds
Channel catfish (US)	*Ictalurus punctatus*	Ponds/streams
Lamprey	*Lampetra fluviatilis*	Fish parasite
Chub	*Leuciscus cephalus*	Swift rivers
Pearl mussel	*Margaritifera margaritifera*	Fast rivers
Signal crayfish	*Pacifastacus leniusculus*	Chalky bottoms
Perch	*Perca fluviatilis*	Mobile in rivers
Roach	*Rutilus rutilus*	Surface feeder
Rainbow trout	*Salmo gairdneri*	Fast streams
Salmon	*Salmo salar*	Fast rivers
Trout	*Salmo trutta*	Fast streams
Rudd	*Scardinius erythrophthalomus*	Mid and surface
Grayling	*Thymallus thymallus*	Clean rivers
Tench	*Tinca tinca*	Slow water

The greater the degree of integration between the water and the drier parts of the garden the greater will be the productivity of the whole.

People Gardens ~

The woman's cause is man's;
they rise or sink
Together.

ALFRED, LORD TENNYSON (1809—92),
THE PRINCESS, 1847

At the outset of this book, we commented on the fact that gardens should be for people. We've talked about the need to include in our designing the needs of all users of the garden, be they young, full of energy and able, elderly and seeking quiet and shade, or wheelchair users with restricted mobility. We stressed the need to think of what people *actually* do in their gardens. Allow for washing, do-it-yourself projects, bicycle storage, garbage/recycling collection and so on.

Even in a small urban garden many people pleasures can be designed in.

Think of your garden full of family and friends and imagine they were all having a good time. Here is a list of things people do (or would like to do) when in my garden:

Sit and read
Play on a climbing frame
Talk in a shady spot
Cook (barbecue)
Pick salads
Pick fruit
Park bicycles
Play ball games
Swing
Eat comfortably
Pick herbs
Grow annual vegetables
Dry washing
Park buggies/prams
Go to the toilet
Store bottles/cans
Do woodworking
Splash in water
Exercise
Get some peace and quiet
Watch the sun go down
Compost refuse
Cut and store firewood
Sunbathe
Store rainwater
Play hide-and-seek
Talk to neighbours
Grow trees

Some readers may be thinking, 'That's all very well but I don't have a garden big enough', or even, 'I don't have a garden at all'. You have to work with what you've got. But let it be said straight away that there's a garden available for everyone. In the UK we have the allotment system where anyone can get access to land for growing food. In addition all cities have waste land begging to be used.

The solution here may be to set up community gardens. A neighbourhood association is formed of interested people who get legal access to the land. This may be temporary or long term. It may be simply for pleasure gardens, or for food production. City farms are examples of sites which combine all these ideas with an educational function. I have seen marvellous examples of community gardens.

In Covent Garden in London there

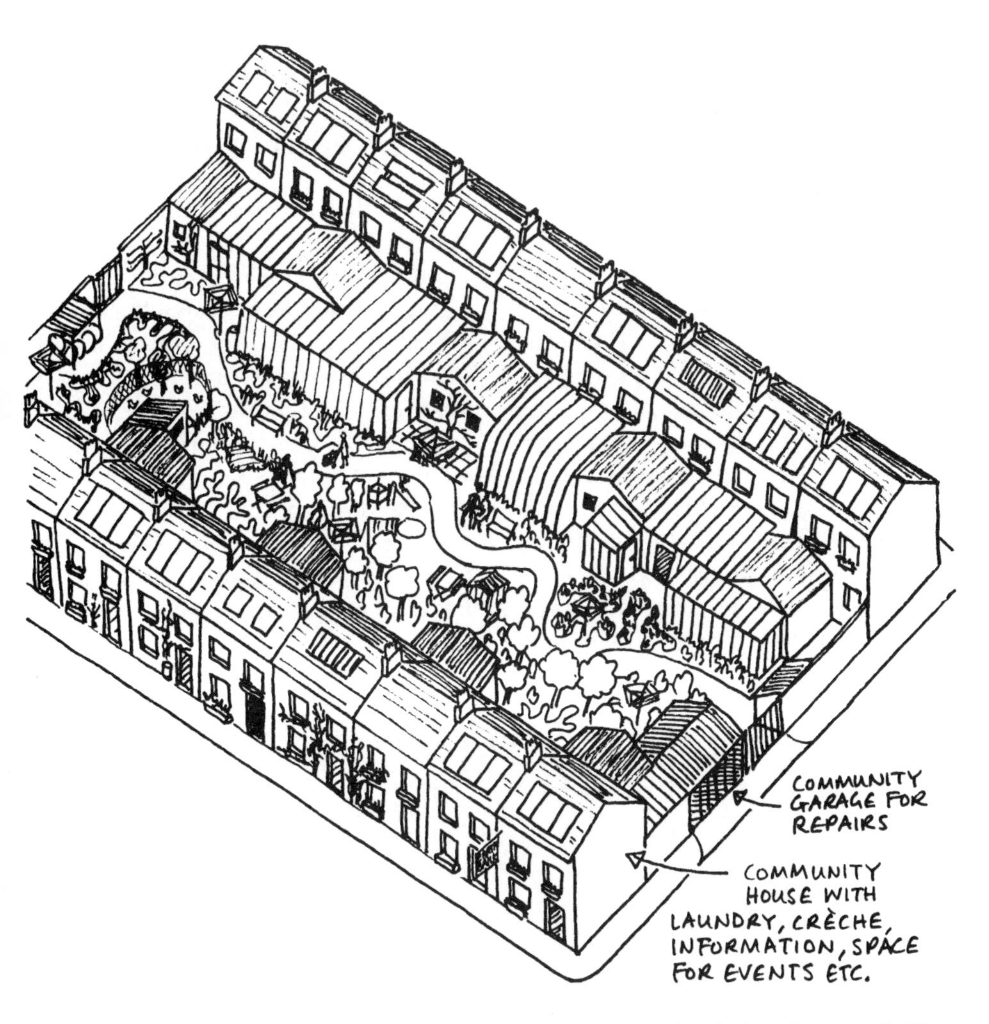

By sharing many small resources a larger one can be built for the use of all.

was a Japanese water garden built by locals which lasted about three years on a temporarily vacant site. It was produced by enthusiastic amateurs, and was delightful for as long as it lasted. What does it matter if it didn't go on forever? People all over the world are learning that unmitigated urban sprawl is hateful and leads to poor quality of life. Recession is an ideal opportunity to re-green those unused urban sites, and fragment the city back into linked villages.

It is also a good time to start producing food for people where they are. Even window boxes can be productive. Flat roofs of tenement blocks can become mulch gardens. Endless expanses of municipal mowed grass can be converted back to food forests. There is nothing to get in your way but lack of ideas, or enthusiasm.

I live in a small town of twelve hundred people. I noticed at the end of one spring weekend that the town dump had increased by at least a tonne of garden refuse. What a waste—this should all be turning into compost and going back on the land. I am inspired to see if we can't set up a community composting scheme. It must be worth a try.

If you feel disempowered by the lack of access to the land in your area, ask yourself how you could turn that about. Inner city alienation reduces people's belief in community action. But it's often 'silly little things' like a community composting scheme that bring people together to appreciate that they have a collective ability to get things done.

The problem of collective working is that we can only reach agreement if we are all willing to let go a little of our own agenda in favour of someone else's. That's a very hard thing to do. Compromise is the name of the game.

Many urban housing areas are built on a block basis with gardens facing inwards. What would happen if folks pulled down the boundaries and shared what they had in common?

It may not work, of course. But then nothing will unless we try. What could be more healing to damaged communities than gardening together?

eleven

Working with Soil

The earth does not belong to man; man belongs to the earth. This we know. All things are connected like the blood which joins one family.

CHIEF SEATTLE, 1854

Our garden must have a place in space and time. The ultimate consideration of a healthy garden, then, is the soil in which we build it. Making the soil fertile will make our plants healthy and abundant.

Plant Needs

Good soil allows plants to root well. This gives them firm support. Good root structure also allows the plant to feed itself well. Growers need to ensure good supplies of water, air and nutrients in the soil. This is done by building soil structure.

A well-structured soil has good 'crumb'—that is, the soil tends to hold together in particles which allow a maximum of 'pore space'. The pores in soil are the spaces between soil particles, and are essential for providing roots with air and water. If the pores are too large water will drain away quickly causing wilting in hot weather. If they are not large enough then the plants can become water-logged in wet weather, or grow slowly because of lack of nutrients.

Water is needed, not just because plants get 'thirsty', but also because it is how plants feed. Most of the minerals they need for growth are absorbed in solution through their root walls. Inside the plant the flow of food from one part to another can only take place if there is enough water to keep the system flowing.

Minerals in the soil come from several sources:

- from the weathered rocks which make up the tiny soil particles
- from humus, or decayed vegetable matter
- from the life cycles of living creatures in the soil
- from the air

The soil type in your district is fixed by history, but you can build the humic content and change the nature of that base material. Humus makes dense soils more porous, and light soils more water retentive. It is a rich source of slowly released plant food. Also you can create conditions which build up soil life and so feed the soil.

Doing both of these will improve the soil structure and bring and keep more air and water where it is available for plant roots to use it. This in turn will help the transport of available minerals to where the plants need them.

Geological Influence

Soil is a living matrix of great complexity. Its origin is in the effects of weather on the landscape and in the subsequent interactions of life with the material created. Over vast periods of time soils build up, and are constantly changing.

Soil is affected by the ground conditions underneath your garden. Each particular bedrock will create different soil as it weathers down. Rock is built up on the earth over time scales which escape human comprehension. Perhaps rocks too are alive, but move too slowly for us to appreciate. They come into being from existing rocks and minerals accumulating, often under great pressure, in a number of ways.

Volcanic rocks are created through eruptions of hot rock from the earth's core. Igneous rocks are the cooled remains of laval flows, ash, and subsurface disturbances of the earth's crust. Granite and pumice are examples of these. Metamorphic rocks are existing rocks which have been changed by the adjacent heat of volcanic activity, or pressure from an overburden. This is how slate is formed. Volcanic rocks, both igneous and metamorphic, tend to be crystalline in nature, and can be very hard.

Sedimentary rocks arise when layers of material are deposited on ocean or river beds, or where windblown material accumulates in deserts. So mudstones and sandstones are born. Over time rocks change and evolve. Young sedimentary rocks may be soft and less dense. Compaction over longer periods may give quite different qualities. Sedimentary rocks tend to be more porous, and to split in a horizontal grain. Upheavals in the earth's crust may change the nature and direction of grain in a rock.

Then comes rain, ice, frost, sun, tide and wind, weathering exposed rocks and carrying particles and pieces to form rocky clusters of mixed materials, or fresh sedimentary layers. In the process living material is trapped to form fossils, or on a grand scale builds coal or chalk–rocks which are entirely made of remnant biomass.

By looking at the existing geobotany we can tell something of the underlying ground. Plants which love salt, acid or poor soils give us an indication of what will be appropriate to plant there.

The underlying geology of your land may have various effects on the garden. It will affect the rate of drainage. Free-draining soils are enjoyed by crops like grapes and carrots. Other plants, such as reeds and rushes, need wetter land.

The chemistry of the rock will affect soil conditions, so that in limestone areas soils will tend to be thinner, yet less acid. Chalk landscapes, for instance, have a very particular natural ecology. On the other hand, in alluvial plains, which form some of the most fertile ground on earth, the overlying material may differ from, and make widely irrelevant, the underlying geology.

Each locality is different, and local knowledge is important. Getting to know the geology of your area helps understanding of your soils, and reconnects you to millions of years of prehistory, which can be a good way to put a bad day into perspective.

There are some easy sources of information. First consult a local geological map, noting the contrast between the ancient geology of the area and any more recent glacial effects. Then look at local soil maps. This information can be compared against the local land use map. The potential of your site becomes much clearer—and you don't have to be an expert geologist to understand this kind of information.

Soil Types

Soil is commonly defined by the nature of its mineral content. The triangular table shows how proportions may vary between pure sand, clay and silt to any admixture of the three.

Take a large jam jar and half fill it

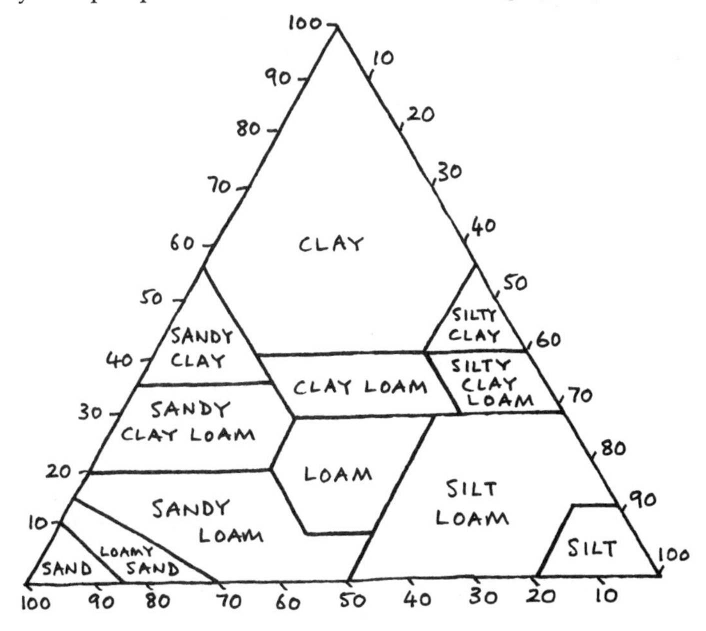

How does your soil fit on this table? Try the jam jar test.

with soil from your garden. Top it up with water the put the lid on. Shake it up and leave it to settle. Largest particles settle quickest, so you get layers up from the bottom of stone, sand, silt, clay. Humus will float. Fine clay may take forty-eight hours to settle out, whilst sand will be pretty instantaneous. The proportion of the layers tells you how your soil is made up.

Sand

Sand is mostly quartz particles, the crystals in eroded rock which are most resistant to being worn down. It has large particle size, and is rich in minerals. However it is hard for plants to use these minerals, and so sand only becomes fertile as soil when it is mixed with silt, clay and/or organic matter. It is free draining, which means there is little problem with wet weather, and sandy soils are 'light' or easy to work any time of year. Root crops (such as carrots) do well in these conditions.

Sandy soils are also common under heath or moorland where long term acidity has leached out the clay fraction of the soil. The gardener's task in these 'laterite' soil conditions is to increase water retention and the availability of nutrients by the introduction of organic matter.

Mulching, and manuring with animal dung or green matter (seaweed is a good one) are the best strategies here. They encourage high populations of earthworms who are the most effective builders of soil structure. This strategy will need to be long term on very sandy soils.

Silt

Silts are midway between clay and sand in grain size, and tend to be rich in a wide range of available minerals. They are typical of the alluvial deposits in river valleys.

Clay

At the other extreme from sand, clay is rich in available minerals, partly because of the small particle size of the material. Clay particles can be one thousand times smaller than sand grains. Clay also has 'colloidal' properties. Colloids are like wallpaper paste—they can hold on to large volumes of water, by a process of very weak electrical attraction. As we've seen, it's not just that water itself is useful in the garden, but that moisture is the means by which many dissolved nutrients travel through the soil in solution.

The danger of clay is that it is poor draining, and so can become waterlogged, starving plants of oxygen and causing roots to rot. Again, organic material will help break up these 'heavy' soils. In addition calcium causes 'flocculation', that is, it makes clay particles stick together, and therefore become more free draining as pore spaces are created between groups of particles. The drawback is that the flocculation is short term. So adding calcium by liming can help improve the texture of clay soils, but the effect is not long lived.

More helpful soils can be built up by trenching rubble and old plaster into the clay, but this is hard work. Better results can be achieved more quickly by building fresh soils over existing clay, rather than by trying to change the soil profile deeper down. In time the plants grown successfully on this soil will initiate the process of soil improvement for you. The soil life will do the rest.

Peat
There are also organic soils, which are very low in mineral content, being composed largely of decayed organic matter. These peat soils tend to acidity and support ranges of plants adapted to such conditions. For cultivating a wide range of conventional crops they will need liming and the addition of mineral content to give available fertility.

The best soils are the loams, which are a good mixture of sand, clay and silt, preferably with high organic content. There are eighteen known elements in soil essential for plant growth:

calcium	sulphur
hydrogen	iron
oxygen	manganese
nitrogen	boron
potassium	molybdenum
phosphorus	copper
carbon	zinc
sodium	chlorine
magnesium	cobalt

These elements need to be present as appropriate compounds combined with other elements to be available to plants. Inappropriate chemical forms mean that good supplies of mineral content can be completely unavailable, and so the soil is effectively infertile.

You can have professional soil samples done which will cost money. The best indicator is plants themselves. If the plants look healthy, then most likely the soil is. Returning all organic wastes to the soil is the best way of insuring the soil stays this way.

Other minor elements are not proven to affect plant yields, but may do so. The proportions in which nutrients are required vary greatly. So an acre crop of maize will take up 68kg (150lb) of nitrogen, but only 10 grammes of boron. Yet if the boron is missing the yield will be adversely affected. Overabundance of certain elements can also 'lock up' others, so for instance we get lime-induced chlorosis in pines on soil which for this tree is too alkaline. The trees suffer, not because there is a lack of nutrient, but because the chemistry of over supply of one nutrient prevents the plant accessing another.

Macro = broad scale, *micro* = smaller scale. Nitrogen, phosphorus and potassium (the N,P,K of chemical fertilizers) are the primary 'macro-nutrients' of soil, and calcium, magnesium and sulphur the secondary ones. The other elements are just as important, but occur in smaller quantities. They may be in short supply in sandy and *organic* soils, organic in this instance meaning soils with low mineral content, such as peat. This is also the case in alkaline soils: not much of a problem in cooler climates where soils tend to acidity, but a threat to plant and human nutrition in tropical and desert conditions.

Acidity and Alkalinity

The 'reaction' of a soil is measured as its acidity or alkalinity. It is registered as its pH reading. A pH of 7.0 shows that a soil is neutral, indicating a balance of hydrogen and hydroxyl ions. Less than 7.0 indicates acid soil and over this figure, alkaline. Don't worry if the chemistry leaves you baffled, you can garden perfectly well without a doctorate in the subject.

The important thing is that the relative acidity or alkalinity of soils affect the availability of nutrients. It's worthwhile testing a cross section of your soils. There are cheap kits or meters available at garden centres to do this yourself, or it's also a service which can be bought from local agricultural colleges etc. The range of pH in your garden may affect your plan, so it's as well to be informed.

Most economically useful plants do well in slightly acid conditions in the average temperate garden. Some plants, such as witch hazel or azalea, definitely need more acid soils to succeed. It is significant, however, that earthworms only become active at a pH over 5.5. So in very acid conditions your friendly slippery soil scientists can't get to work.

The addition of organic matter, such as horse manure or leaves, as soil feed will tend to make the ground more acid; calcareous matter (e.g. lime or chalk) will bring it back to neutral or alkaline. Don't apply both together as the lime will cause the nitrogen feed of the manure to be lost to the atmosphere. The common practice in organic rotations (annual changes of crop to maintain soil health) is to lime before planting cabbage family plants, and to manure well before potatoes.

The Life in the Soil

Well said old mole! Canst work i' the earth so fast?

WILLIAM SHAKESPEARE (1564–1616), *HAMLET*

The mineral aspects of soil chemistry are meaningless unless we look at the *biology* of soil. It's technically possible to grow plants in chemical solutions (known as hydroponics). In nature, however, they exist in an active vibrant soil, with living creatures weighing as much as twenty tonnes per hectare at work below ground. The mineral content of the soil is important, and the physical support of the earth indispensable for the well-being of most plants. Yet it is the life in the earth which actually makes land fertile.

The creatures underground vary from rodents through earthworms, insects, and edible fungi down to bacteria and the *actinomycetes*—strands of microscopic living matter threading through the soil pores. The important function of all these workers is to break down dead and decaying matter into humus. This is a black, colloidal, sponge-like substance which can exist for thousands of years below ground. It has a great attraction for water and thus can become a storehouse for nutrients. Each organism has an important place in relation to others in the food chain, so

Table 24 Plants Tolerant of Acid Soils

All these plants are tolerant of acid soils

Firs	*Abies* spp	P
Alders	*Alnus* spp	P
Bearberry	*Arctostaphylos uva-ursi*	P
Birch	*Betula* spp	P
Heather	*Calluna* spp	P
Common sedge	*Carex nigra*	P
Broom	*Cytisus* spp	P
Heather	*Erica* spp	P
Alpine strawberry	*Fragaria vesca*	P
Gaultheria	*Gaultheria* spp	P
Broom	*Genista* spp	P
Rushes	*Juncus* spp	P
Juniper	*Juniperus communis*	P
Larch	*Larix* spp	P
Liquidamber	*Liquidamber styraciflua*	P
Bog bean	*Menyanthes trifoliata*	P
Bog myrtle	*Myrica gale*	P
Marjoram	*Origanum rotundifolium*	P
Wood sorrel	*Oxalis acetosella*	P
Pines	*Pinus* spp	P
Tormentil	*Potentilla erecta*	P
Primula	*Primula* spp	P
Cloudberry	*Rubus chamaemorus*	P
Willow	*Salix* spp	P
Rowan	*Sorbus aucuparia*	P
Thyme	*Thymus caespititius*	P
Trapaeolum	*Trapaeolum speciosum*	P
Blueberry	*Vaccinium corymbosum*	P
Cranberry (US)	*Vaccinium macrocarpon*	P
Bilberry	*Vaccinium myrtillus*	P
Cowberry	*Vaccinium vitis-idaea*	P

none are just 'pests': all contribute to a complex web of management.

Humus helps granulation of soil particles, making the ground workable and increasing the proportion of pore space in the soil. This in turn makes possible more air and water supplies underground, essential for plant nutrition.

Decaying organic matter supplies

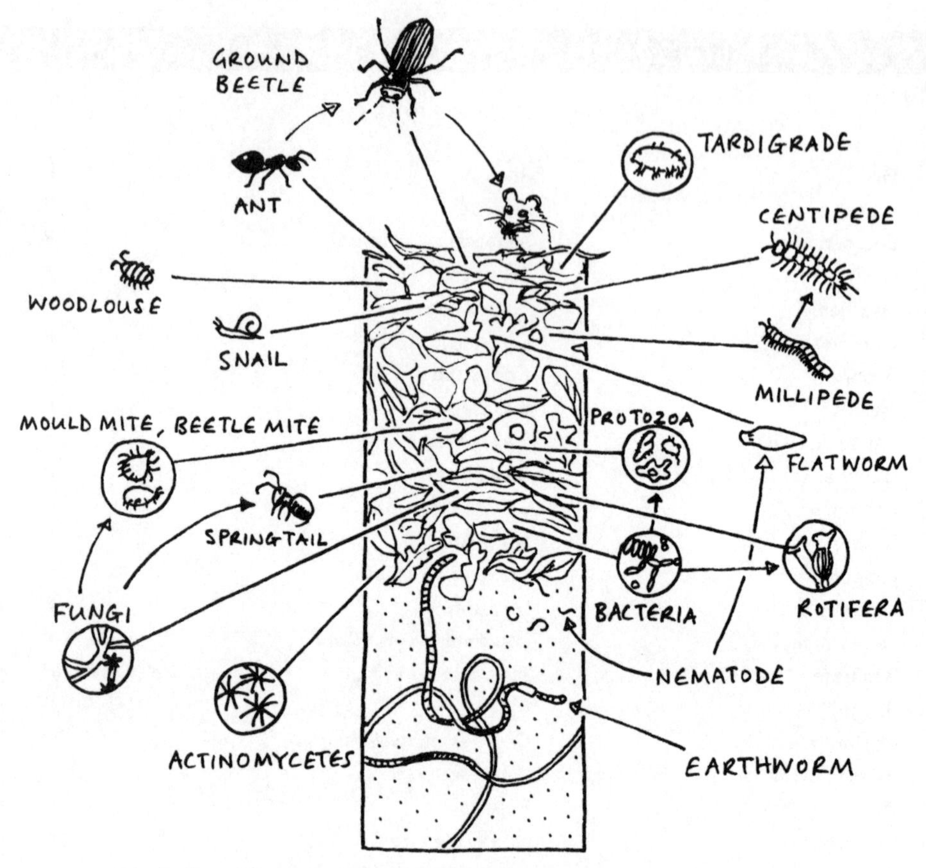

Little bugs have bigger bugs upon their backs that bite 'em.
Bigger bugs have little bugs and so ad infinitum...

nitrogen for plant protein (this is absent in the mineral part of the soil), and also most of the phosphorus and sulphur which plants need.

In turn the humic content is the main food source of soil micro-organisms. So we start to see a picture in which the soil is a teeming city of busy neighbours all living out of each other's pockets.

Plant roots mine the soil, breaking up compaction and enabling water to infiltrate. Animals, birds and insects ingest plant material and leave partially decayed waste on or near the soil surface in their faeces and urine. Earthworms and other burrowing creatures take fresh organic matter from the surface and transport it under the soil. Insects and micro-organisms break up dead plant matter into humus which stores nutrients for the soil life and plants. Bacteria and fungi make food chains between minerals in the soil and living roots. Water levels enable solutions to pass to where they are needed for growth, and so on.

The soil *biota*, this plethora of creatures, is, in summary, all we need to dig the garden. It is also a large self-renewing nitrogen store to feed our crops. Protecting and nurturing it is the proper duty of the gardener.

We mentioned the effects on soil pH of adding lime. Another slow release of calcium can be obtained from calcified seaweed. This has the additional advantage that it is rich in trace elements; indeed, all seaweed products have this benefit. Seaweed should be stored until rainwater has washed out its salt content. For those away from the coast and with deep pockets, liquid seaweed extracts will serve the same purpose. But all of these soil amendments require work and are generally only obtained at a price. Use them, by all means, to establish a plot, but look long term to making the garden self-fertile.

A prerequisite for this is that we return all human wastes to the garden. Many people realize the importance of returning kitchen waste to the soil and usually do so through compost heaps. Better to simply mulch the scraps straight onto the garden: compost heaps get hot, and so do not foster life which will flourish when the residues are returned to the soil. Keep a pile of grass clippings or wood chip handy to cover up untidy looking scraps. Many people already practice such tech-

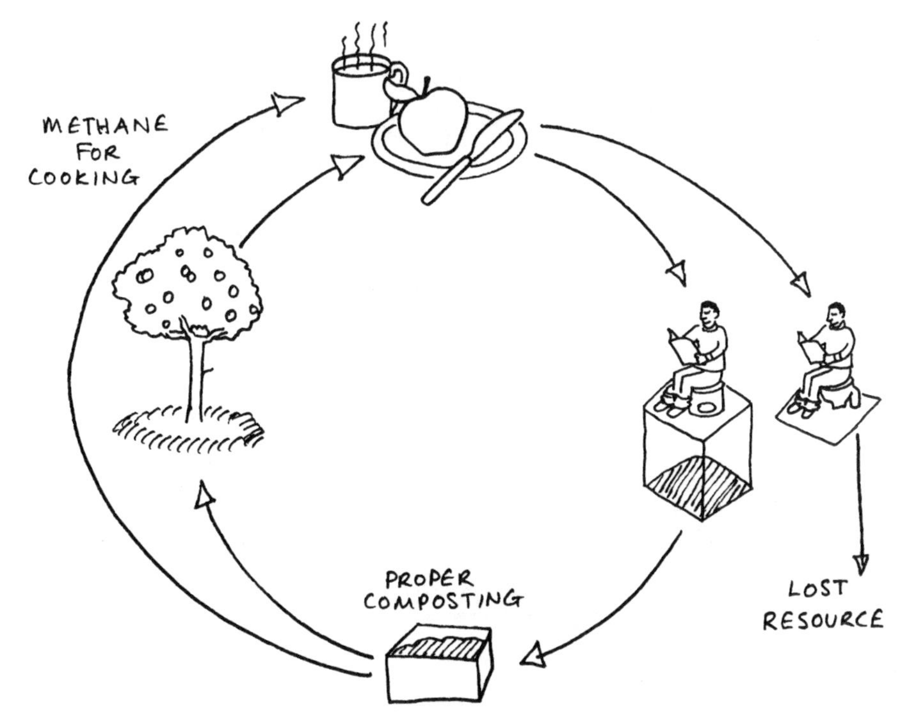

Sustainable gardeners aim to give back what they take.

niques. How many of us, however, put our urine and faeces back on the land?

Victorian standards of hygiene were beneficial in clearing up many nasty illnesses, and today's planning authorities favour the flush toilet over the commode for that reason. Yet for generations the fertility of the cottage garden was assured by the return of night soil to the land. It's obvious isn't it? You can't eat food from the land and throw the excrement somewhere else and expect the land to remain fertile. Properly composted to prevent the cycling of diseases, human waste is an excellent fertilizer.

Amending the Soil

A soil profile can be dug to test the soil. This should be wide enough to stand in, and deep enough to get well into the subsoil. Soils vary greatly within short distances, so the longer the trench, the more you will learn. Be careful to separate the layers of soil and return them in the reverse order from which they came out.

If you get a metre down then you'll learn a lot about what you have to work with. Amending the soil can then be done in the light of solid understanding rather than guesswork. Remember the

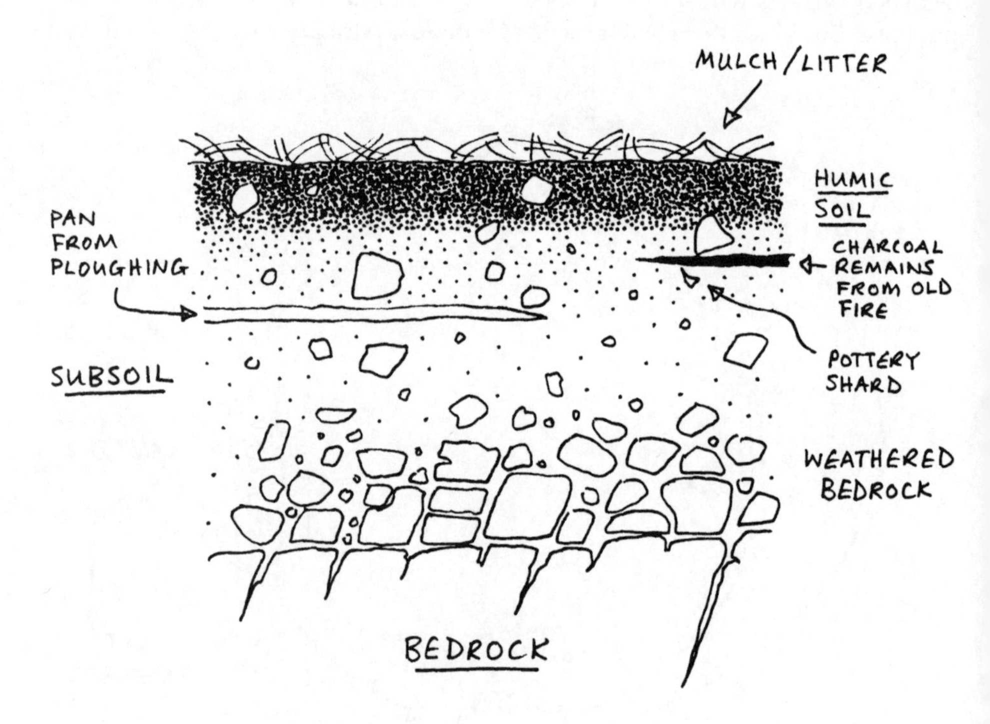

A soil profile tells us the story of our garden soil, helping us plan what will work best..

tools we've already looked at for doing this. Here are some key ones:

- Green manures
- Composting
- Mulching
- Legumes
- Tree planting
- Liming

Systems That Self-Manage Soil ~

The target is to build gardens which are self-managing. The key elements have already been detailed. As a reminder let's summarize them:

- Return all crop wastes to the soil
- Use minimum tillage
- Avoid walking on soil
- Maintain year-round ground cover
- Mulch any bare soil
- Emphasize perennials, especially trees
- Plant communities of species
- Use legumes
- Use green manures
- Weed by suppression rather than digging

Ground Cover ~

Ground cover is particularly important in building good, moist, warm soils. We looked at this in some detail in Chapter 5. Remember to emphasize living cover wherever possible. If that can't be done, use organic mulch in preference to black plastic, which is best used to warm up the underlying soil quickly (see p. **66-7**).

Don't worry if forced to use a less desirable method of ground cover. Plan how to move on to living cover at a later date.

Booklist

Books are the accumulation of historical wisdom preserved for us in this moment, someone clever once said. They are probably the best tool your garden will ever have, for only with understanding do we achieve right action.

Rather than just list books, I've given a one line completely biased review of why I select these texts to recommend. Not many readers will have the time to get through all of these, so I hope the notes will send you in the right direction for your needs for further information.

There are many more good books on the subject than there is space to mention here. New books come out all the time, and if your interests follow particular lines of enquiry, then you may want to look up more specialised works. Many of these books have more detailed references to enable you to do that. Dates are latest editions referred to by the author.

Gardening (General)

Companion Planting 128pp Gertrud Franck (Thorsons, Wellingborough, England, 1983)
A comprehensive system based around green manures and plant relationships. Useful ideas.

Designing and Maintaining Your Edible Landscape Naturally 370pp Robert Kourik (Metamorphic Press, Santa Rosa, California, 1984)
The gardening book this author wishes he'd written. Detailed science, and great visuals. Excellent source book for North American gardeners.

The Green Gardener's Handbook–everything you need to know for successful organic growing 336pp Margaret Elphinstone and Julia Langley (Thorsons, London, 1990)
General survey of organic growing. Splendid species list of common garden plants and their maintenance. Ignore the bit about bleaching your patio.

Harvests & Harvesters 222pp John Hargreaves (Victor Gollancz Ltd, London, 1987)
An excellent work that deserves to be more widely read studying the

production of various food crops in Britain. Honestly appraises the real challenges for those who feed the nation.

New Plants from Old 116pp Charles M.Evans (Pelham Books, London, 1976)
A good manual on do it yourself propogation.

The No-Work Garden Book 188pp Ruth Stout and Richard Clemence (White Lion Publishers, London, 1976; originally Rodale Press USA, 1971)
Ruth Stout's method comprehensively explained.

Organic Gardening 240pp Lawrence D. Hills (Penguin, London, 1977)
Conventional organics.

Organic Gardening: Pocket Encyclopaedia 224pp Ed. Geoff Hamiliton and others (Dorling Kindersley, London, 1991)
More simple and more pictures than Hills.

The Salad Garden 168pp Joy Larkcom (Windward, Leicester, 1987)
Beautifully illustrated. An irresistable temptation to grow and eat salads.

Veganic Gardening 144pp Kenneth Dalziel O'Brien (Thorsons, Northants, 1986)
Good logical exposition–even if you're not vegan.

The Vegetable Garden Displayed 148pp Joy Larkcom (Royal Horticultural Society, Surrey, 1981)
A good comprehensive manual.

The Victorian Kitchen Garden 160pp Jennifer Davies (BBC Books, London 1987)
Not 100 per cent accurate but most enjoyable. Victorian inventiveness leaves us standing!

Weeds. How to Control and Love Them 64pp Jo Readman (HDRA/Search Press, Kent 1991)
An excellent, well-illustrated short explanation of what weeds are and what they do for us.

Food

Alternative Foods 177pp James Sholto Douglas (Pelham Books, London, 1978)
A masterly work on the possibilities for extending small-scale food production to wider realms.

Food & Society 178pp Magnus Pyke (John Murray, London, 1968)
Scholarly and a fun read at the same time.

Food for Free 240pp Richard Mabey (Collins, London 1989)
The easiest to follow guide for wild food gatherers. Great pictures.

Raw Energy 315pp Leslie and Susannah Kenton (Arrow Books, London, 1989)
Getting well through careful eating.

They Can't Ration These 123pp Vicomte de Mauduit (Michael Joseph, London, 1940)
Wartime necessity encourages out old knowledge.

Fruit

The Berry Garden 255pp Mary Forsell (Macdonald Orbis/Running Heads Incorporated, London/New York, 1989)
Not all berries are edible. Very beautiful book extending the range of possibilities, particularly for shrubs.

Cultivated Fruits of Britain 349pp F.A. Roach (Basil Blackwell, Oxford, 1985)
Fascinating history of fruit development.

The Fruit Garden Displayed 223pp Harry Baker (Cassell Ltd/The Royal Horticultural Society, London, 1986)
Sober and exhaustive manual.

The Good Fruit Guide 81pp Lawrence Hills (Henry Doubleday, Essex, 1984)
A more enthusiastic go at the same subject. Not so well illustrated.

The Grafter's Handbook 323pp R.J. Garner (Cassell, London, 1988)
Exhaustive and fascinating. A great encouragement to have a go in an area where many lack the confidence to try.

Health, Herbals and Uses of Plants

Edible & Medicinal Plants 288pp Edmund Launert (Hamlyn, London, 1989)
Very detailed and illustrated.

Living More With Less 294pp Doris Janzen Longacre (Herald Press, Scotdale, Pennsylvania, USA, 1980)
A pleasant little homily on the morality of it all.

Pests, Diseases & Disorders of Garden Plants 512pp Stefan Buczacki and Keith Harris (Collins, London, 1981)
Extremely comprehensive and terrifying descriptions of what can go wrong. Essential for plant hypochondriacs.

The Role of Medicine 207pp Thomas McKeown (Basil Blackwell, Oxford, 1979)
Puts the politics of health and food in context.

Livestock

Backyard Dairy Book 111pp Andrew Singer and Len Street (Prism Press, Bridport, England, 1978)

Backyard Fish Farming 170pp Paul Bryant, Kim Jauncey, Tim Atack (Prism Press, Bridport, England, 1986)

Ducks and Geese at Home 51pp Michael and Victoria Roberts (Domestic Fowl Trust, Evesham, England, 1985)
All good small-scale manuals for the beginner.

Permaculture and Appropriate Technology

Another Kind of Garden 58pp Ida and Jean Pain (Self published, Villecroze, France, 1972)
The use of biomass for energy and fertility creation.

A Chinese Biogas Manual 135pp translated by Michael Crook, edited by Ariane van Buren (Intermediate Technology Publications, London, 1983)
Community-scale biomass and energy systems.

The Field Director's Handbook 512pp ed. Pratt & Boyden (OUP for Oxfam, 1985)
An invaluable manual for development workers in rural regeneration.

Permaculture One 127pp Bill Mollison, David Holmgren (Tagari Press, Tyalgum, Australia, 1990)

Permaculture Two 150pp Bill Mollison (Tagari Press, Tyalgum, Australia, 1979)

Permaculture–a Designer's Manual 576pp Bill Mollison (Tagari Publications, Tyalgum, Australia, 1988)
Three from the originator of Permaculture. Dry wit, humane creativity, and encyclopaedic knowledge.

The Permaculture Plot 52pp edited by Simon Pratt (Permaculture British Isles, Buckfastleigh, England, 1991)
Some places to see sustainable projects at work.

The Permaculture Way 240pp Graham Bell (Thorsons, London, 1992)
Sustainability in a wider social context. An introduction.

Radical Technology 304pp edited by Godfrey Boyle, Peter Harper (Wildwood House, London, 1976)
A delightful patchwork of creative ideas.

The Timeless Way of Building 550pp Christopher Alexander (Oxford University Press, New York, 1979)
A prose poem in itself. Recommended start point for all designers.

Also: the practical series from British Trust for Conservation Volunteers, 36 St Mary's Street, Wallingford, Oxfordshire, England, OX10 0EU, are invaluable manuals on 'how to':
Drystone Walling 120pp 1986
Fencing 141pp 1986
Footpaths 192pp 1983
Woodlands 173pp 1988

Much 'back to the land' work in the UK was originally inspired by the books of John (and originally Sally) Seymour. Still worth referring to is:

The Forgotten Arts 192pp John Seymour (Dorling Kindersley, London 1986)
A beautifully illustrated record of how many household artefacts can be made by hand from locally available materials.

Crofting Years 142pp Francis Thompson (Luath Press, Barr, Scotland, 1988)
An insight into the importance of growing to the cohesion of small communities in the most difficult circumstances.

Growth of the Soil 317pp Knut Hamsun (Picador, London, 1980)
Whilst not mentioned in the present work, a wonderful read to show the kind of connection with the land which can be made through growing.

Philosophy, Poetry and Essays

In Search of Our Mothers' Gardens 397pp Alice Walker (The Women's Press, London, 1984; originally Harcourt Brace Jovanovitch, New York, 1983)

Selected Poems 160pp D.H. Lawrence (Penguin, Middlesex, 1950)

Thinking Like A Mountain 122pp Seed, Macy, Fleming and Naess (New Society Publishers, Philadelphia, 1988)

Soil, Climate, Geology

The Formation of Vegetable Mould (Through the Action of Worms With Observations on their Habits) 152pp Charles Darwin, Introduction by Sir A. Howard (Faber & Faber, London, 1945)
If you can get hold of a copy, Darwin's greatest and most neglected work.

Gaia: A new look at life on Earth 154pp J.E. Lovelock (Oxford University Press, 1987)
Our planet as a living organism. Puts irritating people into perspective.

Geological Structures 250pp John L. Roberts (Macmillan Field Guides, London, 1989)

Meteorology–The Atmosphere and the Science of Weather 502pp Joseph M. Moran and Michael D. Morgan (Macmillan, New York, 1986)
Two easy field guides.

The Nature & Properties of Soils 8th Edition 639pp Nyle C. Brady (Macmillan, New York, 1974)
The university text book of soil science. Well written and intelligible to comparatively lay folk. Surpri- singly comes down strongly against tillage!

Trees

The Forest Garden 24pp Robert Hart (Institute for Social Inventions, London, 1990)
Simple but marvellously comprehensive guide to the idea.

Forest Gardening 212pp Robert Hart (Green Books, Devon, 1991)
A longer book, more easily available, but not as crisp.

The Hillier Colour Dictionary of Trees and Shrubs 323pp Hillier Nurseries (David & Charles, Newton Abbot, England 1988)
Easily available, good photos.

Silviculture of Broadleaved Woodlands 232pp Julian Evans (HMSO [Forestry Commission], London, 1984)
Professional manual, accessible to the amateur. Not in favour of mono-culture excess.

Soil Care and Management 48pp Jo Readman (HDRA/Search Press, Kent, 1991)
Clear illustration of soil processes and management techniques.

Trees in Britain Europe & North America 223pp Roger Phillips (Pan, London, 1978)
Good visual guide. Comprehensive list, but little detail.

Trees, Woods and Man 272pp H.L. Edlin (Collins, London, 1978)
Social impact of trees.

Species

Agar to Zenry 152pp Ron Freethy (The Crowood Press, Marlborough, England, 1985)
Ron Freethy is a credit to British naturalists, and is not afraid to see the value of 'homely' knowledge, based as it is on generations of experience.

The Book of Bamboo 340pp David Farrelly (Sierra Club Books, San Francisco, 1984)
Brilliant. You never knew it was such useful stuff!

British Plant Communities: Vol.1 395pp ed J.S. Rodwell (Cambridge University Press, 1991)
A scientific work of gargantuan research for those who seriously want to take up the subject.

British Plants 152pp H.L.Edlin (B.T.Batsford, London 1951)

Grasses, Ferns, Mosses & Lichens 191pp Roger Phillips (Pan Books, London, 1980)
Good field guide.

The International Permaculture Species Yearbook 144pp Edited by Dan Hemenway (Self published, Orange, Ma, USA, 1986)
Lots of dense type and no pictures with detailed global information. One of a series.

Mushrooms in the Garden 152pp Hellmut Steineck (Mad River Press, Eureka California, 1984; originally *Pilze im Garten* Eugen Ulmer, Stuttgart, 1981)
How to do it.

Mushrooms 288pp Roger Phillips (Pan Books, London, 1985)
What they look like from our greatest faunal photographer.

A Passion for Mushrooms 192pp Antonio Carluccio (Pavilion Books, London, 1990)
As with all the best writers, the passion is not only informative on finding and cooking, but the book reads like poetry, with excellent pictures by Roger Phillips and others.

Plant Communities 128pp Anne Bulow-Olsen (Penguin, Middlesex, 1978)
The only book for the lay reader I've found which explains the principles (with twenty examples).

Plants & Beekeeping 236pp F.N. Howes (Faber & Faber, London, 1979)
Easy-to-use advice on what feeds our honeyed friends.

Further Contacts

Australia

Permaculture International Journal, PO Box 7185, Lismore Heights, NSW 2480.

Permaculture Western Australia, Warwick Rowell, PO Box 148, Inglewood, WA 6052.

Germany

Harald Wedig, Permaculture Institut, Bruchstrasse 94, D-41749, Viersen.

New Zealand

Robin McCurdy, Tui Community, Wainui Inlet, RDI, Takaka, Aotearoe.

Russia

Vlodya Shestakov, 64 Gorskhovoaya Street, Apartment 7, PO Box 644, St Petersburg, 191180.

UK

Permaculture UK, PO Box 1, Buckfastleigh, Devon, TQ11 0LF.

USA

Permaculture Activist, Route 1, Box 38, Primm Springs, Tennessee, 38476.

Details of many other contacts can be obtained from these addresses.

Index